London Mathematical Society Student Texts

Managing editor: Professor J. W. Bruce, Department of Mathematics
University of Liverpool, Liverpool L69 3BX, United Kingdom

Titles are available from http://www.cambridge.org

London Mathematical Society Student Texts 60

Linear Operators and Linear Systems
An Analytical Approach to Control Theory

JONATHAN R. PARTINGTON
University of Leeds

CAMBRIDGE
UNIVERSITY PRESS

PUBLISHED BY THE PRESS SYNDICATE OF THE UNIVERSITY OF CAMBRIDGE
The Pitt Building, Trumpington Street, Cambridge, United Kingdom

CAMBRIDGE UNIVERSITY PRESS
The Edinburgh Building, Cambridge CB2 2RU, UK
40 West 20th Street, New York, NY 10011-4211, USA
477 Williamstown Road, Port Melbourne, VIC 3207, Australia
Ruiz de Alarcón 13, 28014 Madrid, Spain
Dock House, The Waterfront, Cape Town 8001, South Africa

http://www.cambridge.org

© Jonathan R. Partington 2004

First published 2004

Printed in the United States of America

Typeface Computer Modern 9.5/12 pt.   *System* LaTeX 2_ε [AU]

A catalog record for this book is available from the British Library.

Library of Congress Cataloging in Publication data

Partington, Jonathan R. (Jonathan Richard), 1955–
Linear operators and linear systems : an analytical approach to control theory / Jonathan
R. Partington.
p. cm. — (London Mathematical Society student texts ; 60)
Includes bibliographical references and index.
ISBN 0-521-83734-0 – ISBN 0-521-54619-2 (pbk.)
1. Linear operators. 2. Linear systems. I. Title. II. Series.
QA329.2.P37 2003
515′.7246—dc22 2003066665

ISBN 0 521 83734 0 hardback
ISBN 0 521 54619 2 paperback

Contents

Contents

Preface

It should be emphasised at the start that this book does not claim to be an exhaustive treatise on either linear operators or linear systems, but it presents an introduction to the common ground between the two subjects, one pure mathematical and one applied, by regarding a linear system as a (causal) shift-invariant operator on a Hilbert space such as $\ell^2(\mathbb{Z}_+)$ or $L^2(0, \infty)$. It therefore includes material on Hardy spaces, shift-invariant operators, the commutant lifting theorem, and almost-periodic functions, which might traditionally be regarded as "pure" mathematics, and is suitable for those working in analysis who wish to learn more advanced material on linear operators.

At the same time, it is hoped that students and researchers in systems and control will find the approach taken attractive, including as it does much recent material on the mathematical side of systems theory, which cannot easily be found elsewhere: these include recent developments in robust control, power signal spaces, and the input–output approach to time-delay systems. Parts of this book have been expounded in graduate courses and other lectures at that level and could be used for a similar purpose elsewhere.

Chapter 1 begins with a review of basic operator theory without proofs. All this material can be found in any introductory course and many textbooks, and so is included mostly for reference. The other main topic of this chapter, which is treated in considerably more detail, is that of Hardy spaces, which are Banach spaces of analytic functions on the disc or half-plane. Our treatment covers the essential ideas (in particular inner and outer functions) that will be needed later.

In Chapter 2 we begin with material that will be unfamiliar to many readers, namely, the study of unbounded closed operators. The approach is to study an operator by means of its graph, and we also introduce semigroups and the gap metric. Most of the material is fairly standard, although its approach is slanted towards the applications to be encountered later. We also include a brief discussion of admissibility, a topic of active research that has not yet reached the standard textbooks.

Chapter 3 establishes the link between linear systems and shift-invariant operators. The standard Beurling–Helson–Lax–Wiener theorems on invariant subspaces are presented, with the most elementary proofs available; then we go on to shift-invariant operators, which can be thought of as operators with shift-invariant graphs. The other property that distinguishes a linear system is the idea of causality, and we include a discussion of the now-notorious Georgiou–Smith paradox. We conclude with a gentle treatment of the commutant lifting theorem, an abstract theorem in operator theory with many attractive applications.

Chapter 4 brings together the ideas of the previous chapter to discuss robust control (i.e., stabilization by feedback) from a graph point of view using coprime factorizations. This is by no means a new idea, but the mathematical developments of the 1990s that link it with the idea of shift-invariance have put this on a much more rigorous footing, and it is time to present the material in a more operator-theoretic way. We also discuss the chordal metric, originating in complex analysis, which under the name "pointwise gap metric" provides another useful way of measuring the distance between linear systems.

Chapter 5 presents the topic of spaces of persistent signals. These interest engineers greatly, but the subject is a minefield, in that several errors are reproduced in the literature. We start conventionally with an easy introduction to almost-periodic functions, presented as the simplest persistent signals; there are many accounts of these, but no recent ones seem to be as clear as the original ones of Bohr and Besicovitch; these in turn suffer from using what is now a rather old-fashioned notation. After that we move on to bounded-power signals, where we draw largely on papers written within the last five years. We conclude with non-stochastic approaches to white noise, correlation, and parts of Wiener's generalized harmonic analysis, all treated from an operator-theory viewpoint.

Chapter 6 begins with a brief discussion of finite-dimensional systems, but the main topic is that of delay systems, which we may think of as the simplest and most important example of an infinite-dimensional system, having the greatest interest from an analytic point of view. The four themes of this chapter are the classification of delay systems into retarded, neutral and advanced types; stability, a question that resolves itself into asking when delay systems represent operators of multiplication by bounded functions, and how to locate the poles of their transfer functions; rational approximation, which can now be presented in an elementary fashion using shift operators; and finally stabilization, where we apply the ideas of Chapter 4 to this concrete situation. In all cases we give a presentation with the minimum of unnecessary technical detail.

The book includes approximately 100 exercises, which are mostly intended to be quite easy and to give further practical illustrations of the main results in the text.

Since this book's theme is the link between operator theory and systems theory, we deliberately omit certain topics that do not fit into this approach, for example, the spectral theory of non-compact operators. The reader wishing to learn more advanced functional analysis and operator theory is recommended to consult the books [25, 67, 117, 146], for example. Likewise, most treatments of systems theory tend to concentrate on the finite-dimensional case and then go into more details of control design. Of these, the texts [37, 49, 66, 75, 132] may be recommended.

We take the opportunity to mention some other interesting books that take an operator-theoretic approach to linear systems, for example, [31, 32, 38]. These seem to be pitched at a more advanced level, and their selection of material is rather different. This last comment is appropriate for the monographs [20, 151], which we may expect to become classics.

I am grateful to Birgit Jacob, Romesh Kumar, Pertti Mäkilä, Denis Matignon and Martin Smith for their helpful comments on the first draft of this manuscript. Some of the book was written when I was visiting the University of Lyon I, and I thank Isabelle Chalendar and Monique Gaffier for their help in making this possible. I also wish to thank Roger Astley and Elise Oranges for their advice on the production of this book. Finally, this book is dedicated affectionately to Andrew, Chris and Kate, who between them have put up with me for more than a century.

Chapter 1

Operators and Hardy spaces

In this chapter we begin by reviewing the main definitions and theorems from the basic theory of linear operators that will be needed. This material is very standard and likely to have been met in any basic course on functional analysis, and so we give just the essentials of the subject, without proofs.

We then move on to a discussion of the Hardy spaces, which are Banach spaces of functions that can be defined either in the unit disc $\mathbb{D}$ or the right half-plane $\mathbb{C}_+$ and extended, respectively, to the unit circle $\mathbb{T}$ or the imaginary axis $i\mathbb{R}$. Here we give a fairly elementary account of those parts of the subject that are most useful in applications, including the concepts of inner and outer functions. There are many more detailed treatments available, and we refer the interested reader to the notes at the end of the chapter.

1.1 Banach spaces and bounded operators

We shall work with normed spaces, which can be real or complex, and write $\mathbb{K}$ for the field of scalars ($\mathbb{R}$ or $\mathbb{C}$) when it is not important which we take. A complete normed space is called a *Banach space*. An inner product on a vector space induces a norm by means of the formula $\|x\| = \langle x, x\rangle^{1/2}$, and a complete inner-product space is called a *Hilbert space*.

A *linear operator* T from a normed space $\mathcal{X}$ to a normed space $\mathcal{Y}$ is just a linear mapping, that is, it satisfies

$$T(a_1 x_1 + a_2 x_2) = a_1 T x_1 + a_2 T x_2 \quad \text{for all } x_1, x_2 \in \mathcal{X} \quad \text{and } a_1, a_2 \in \mathbb{K}.$$

The operator T is said to be *bounded*, if there is a constant $K > 0$ such that $\|Tx\| \leq K\|x\|$ for all vectors x in $\mathcal{X}$. The least such K that holds for all x is the *norm* of T, written

$$\|T\| = \sup_{x \neq 0} \frac{\|Tx\|}{\|x\|} = \sup_{\|x\|=1} \|Tx\|,$$

1

where the last equality holds because $\|Tx\|/\|x\| = \|T(x/\|x\|)\|$ and $x/\|x\|$ is a vector of norm 1. The boundedness condition implies that $\|Tx - Ty\| \leq K\|x - y\|$ for all $x, y \in \mathcal{X}$, and hence T is continuous. Conversely, continuous operators are always bounded (see the exercises).

The bounded operators from $\mathcal{X}$ to $\mathcal{Y}$ form a normed space, with the norm defined as above, which we shall denote by $\mathcal{L}(\mathcal{X}, \mathcal{Y})$. If $\mathcal{Y}$ is a Banach space, then so is $\mathcal{L}(\mathcal{X}, \mathcal{Y})$. Two special cases are of interest:

1. We denote $\mathcal{L}(\mathcal{X}, \mathcal{X})$ by $\mathcal{L}(\mathcal{X})$. Apart from being a normed space, this is also an algebra, since we can define the product ST of two operators S and T by $(ST)(x) = S(Tx)$ for $x \in \mathcal{X}$; then $\|ST\| \leq \|S\| \|T\|$.

2. The space $\mathcal{L}(\mathcal{X}, \mathbb{K})$ is the *dual space* of $\mathcal{X}$, denoted by $\mathcal{X}^*$. Its elements are the *linear functionals* on $\mathcal{X}$.

In particular, the dual space of a Hilbert space $\mathcal{H}$ can be identified with the space itself, because every linear functional $f : \mathcal{H} \to \mathbb{K}$ is given by the formula $f(x) = \langle x, y \rangle$ for some unique $y \in \mathcal{H}$. Moreover, $\|f\| = \|y\|$. This is the *Riesz–Fréchet theorem*.

Now let $\mathcal{X}$ be a complex Banach space. For $T \in \mathcal{L}(\mathcal{X})$, the *spectrum* of T is the set

$$\sigma(T) = \{\lambda \in \mathbb{C} : T - \lambda I \text{ is not invertible}\}.$$

It is known that $\sigma(T)$ is a non-empty compact subset of $\mathbb{C}$ and that, letting $r(T)$ denote the *spectral radius*, $\sup\{|\lambda| : \lambda \in \sigma(T)\}$, we have

$$r(T) = \lim_{n \to \infty} \|T^n\|^{1/n} = \inf\{\|T^n\|^{1/n} : n \geq 1\}.$$

In particular, $r(T) \leq \|T\|$. We also have $\sigma(T) \supseteq \sigma_p(T)$, where $\sigma_p(T)$ denotes the point spectrum of T, the set of eigenvalues of T. If $\mathcal{X}$ is finite-dimensional, these two sets coincide and are finite and non-empty; however, in the infinite-dimensional case, they can be very different (see the exercises).

The *resolvent set*, $\rho(T)$, is the complement of $\sigma(T)$ in $\mathbb{C}$, that is, the set of points $\lambda \in \mathbb{C}$ for which $(T - \lambda I)^{-1}$ exists. We also refer to $(T - \lambda I)^{-1}$ as the *resolvent* of T, which can be regarded as an operator-valued function $R : \rho(T) \to \mathcal{L}(\mathcal{X})$, with $R(\lambda) = (T - \lambda I)^{-1}$.

For a bounded operator $T : \mathcal{H} \to \mathcal{K}$ between Hilbert spaces $\mathcal{H}$ and $\mathcal{K}$, the *adjoint* $T^* : \mathcal{K} \to \mathcal{H}$ is defined by the equation

$$\langle Th, k \rangle = \langle h, T^*k \rangle \qquad \text{for all } h \in \mathcal{H} \text{ and } k \in \mathcal{K}. \tag{1.1}$$

The following properties are well known and not difficult to prove. They hold for all T, T_1 and T_2 in $\mathcal{L}(\mathcal{H}, \mathcal{K})$ and a_1, a_2 in $\mathbb{C}$:

- $(a_1T_1 + a_2T_2)^* = \overline{a_1}T_1^* + \overline{a_2}T_2^*$;

- $(T^*)^* = T$;

- $\|T^*\| = \|T\|$;

- $(T_1T_2)^* = T_2^*T_1^*$.

Now three special classes of operator are of interest to us:

1. The operator T is *Hermitian*, or *self-adjoint*, if $T = T^*$.

2. T is *unitary*, if $T^* = T^{-1}$, that is, $TT^* = T^*T = I$.

3. T is *normal*, if $TT^* = T^*T$. Clearly both Hermitian and unitary operators are normal.

If T is Hermitian, then $\sigma(T) \subset \mathbb{R}$; whereas, if T is unitary, then $\sigma(T) \subseteq \mathbb{T} = \{z \in \mathbb{C} : |z| = 1\}$.

Suppose that $\mathcal{K}$ is a closed subspace of a Hilbert space $\mathcal{H}$. Then $\mathcal{H}$ has an orthogonal decomposition $\mathcal{H} = \mathcal{K} \oplus \mathcal{K}^\perp$, where $\mathcal{K}^\perp$, the *orthogonal complement* of $\mathcal{K}$, is the closed subspace $\mathcal{K}^\perp = \{x \in \mathcal{H} : \langle x, k \rangle = 0 \text{ for all } k \in \mathcal{K}\}$. Thus every vector $y \in \mathcal{H}$ decomposes uniquely as $y = k + k'$, where $k \in \mathcal{K}$ and $k' \in \mathcal{K}^\perp$, and one has $\|y\|^2 = \|k\|^2 + \|k'\|^2$.

If we now define $P_\mathcal{K} : \mathcal{H} \to \mathcal{H}$ by $P(k + k') = k$, then $P_\mathcal{K}$ is a linear operator on $\mathcal{H}$, the *orthogonal projection* onto K, and it satisfies $P_\mathcal{K} = P_\mathcal{K}^*$, $P_\mathcal{K} = P_\mathcal{K}^2$ and $\|P_\mathcal{K}\| = 1$ (unless $\mathcal{K} = \{0\}$, when of course $P_\mathcal{K}$ is the zero operator). Moreover, $P_\mathcal{K} + P_{\mathcal{K}^\perp} = I$, the identity operator.

An operator $T \in \mathcal{L}(\mathcal{X}, \mathcal{Y})$ between Banach spaces is said to be *compact* if T maps bounded subsets of $\mathcal{X}$ into relatively compact subsets of $\mathcal{Y}$ (that is, sets with compact closure). In particular, finite-rank operators are compact. Equivalently, T is compact if, whenever (x_n) is a bounded sequence in $\mathcal{X}$, the sequence (Tx_n) has a convergent subsequence in $\mathcal{Y}$.

The spectrum of a compact operator is particularly simple. It consists of a finite or countably infinite number of points; and if there are infinitely many, they form a sequence tending to zero. All non-zero points of the spectrum are eigenvalues, and the eigenspaces are finite-dimensional.

Suppose now that $T \in \mathcal{L}(\mathcal{H})$, where $\mathcal{H}$ is a Hilbert space. If T is both compact and normal, then T can be decomposed in terms of its non-zero eigenvalues (λ_k)

and eigenvectors (e_k), which can be taken to be an orthonormal sequence as follows:

$$Tx = \sum_k \lambda_k \langle x, e_k \rangle e_k, \qquad (x \in \mathcal{H}).$$

The eigenvalues tend to zero if there are infinitely many of them. Indeed, all operators of this form are both compact and normal.

A general compact operator $T \in \mathcal{L}(\mathcal{H})$ has the following *singular value decomposition*;

$$Tx = \sum_k \sigma_k \langle x, e_k \rangle f_k, \qquad (x \in \mathcal{H}), \tag{1.2}$$

where now (e_k) and (f_k) are two orthonormal sequences, possibly finite, the *Schmidt pairs* of T, and the constants σ_k are positive real numbers, the *singular values* of T, and form a decreasing sequence: if there are infinitely many, they tend to zero. Indeed, (σ_k^2) are the eigenvalues of the compact operator T^*T. In particular, every compact operator on a Hilbert space is the norm limit of a sequence of finite-rank operators – just truncate the sum in (1.2).

1.2   Hardy spaces on the disc and half-plane

We begin by defining the Hardy spaces as Banach spaces of analytic functions on the disc $\mathbb{D} = \{z \in \mathbb{C} : |z| < 1\}$ and then see them in another light as spaces of functions defined on the unit circle $\mathbb{T} = \{z \in \mathbb{C} : |z| = 1\}$, which we equip with normalized Lebesgue measure.

Definition 1.2.1 *For $1 \le p < \infty$, the Hardy space H^p is defined as the space of all analytic functions f in the disc $\mathbb{D}$ for which the norm*

$$\|f\|_p = \sup_{r<1} \left(\frac{1}{2\pi} \int_0^{2\pi} |f(re^{i\omega})|^p \, d\omega \right)^{1/p}$$

is finite. The space H^∞ consists of all bounded analytic functions f in the disc with norm

$$\|f\|_\infty = \sup_{|z|<1} |f(z)|.$$

It is not hard to see that for $p \le q$ we have $H^p \supseteq H^q$; thus $H^\infty \subseteq H^2 \subseteq H^1$.

The following result holds, although we shall not require its full power and will prove slightly simpler results for $p = 2$ and $p = \infty$.

Theorem 1.2.2 *For functions f in H^p with $1 \le p \le \infty$, the radial limit*

$$\tilde{f}(e^{i\omega}) = \lim_{r \to 1} f(re^{i\omega})$$

exists almost everywhere in t, and indeed $\tilde{f} \in L^p(\mathbb{T})$, with $\|f\|_{H^p} = \|\tilde{f}\|_{L^p}$.

We normally identify f with $\tilde{f}$ and can thus regard H^p as a closed subspace of $L^p(\mathbb{T})$, and hence a Banach space.

It is also possible to start by defining H^p directly as the subspace of those $L^p(\mathbb{T})$ functions for which the negative Fourier coefficients vanish, that is,

$$\hat{f}(n) = \frac{1}{2\pi} \int_0^{2\pi} \tilde{f}(e^{i\omega}) e^{-in\omega} \, d\omega = 0 \qquad (n < 0).$$

Then a function $\tilde{f}$ with $\tilde{f}(e^{i\omega}) \sim \sum_{n=0}^{\infty} \hat{f}(n) e^{in\omega}$ can be naturally identified with the power series $f(z) = \sum_{n=0}^{\infty} \hat{f}(n) z^n$, defining an analytic function f in $\mathbb{D}$.

Proof of part of Theorem 1.2.2: We begin with $p = 2$. For a function $f : z \mapsto \sum_{n=0}^{\infty} a_n z^n$, it is easily verified that we have

$$\frac{1}{2\pi} \int_0^{2\pi} |f(re^{i\omega})|^2 \, d\omega = \sum_{n=0}^{\infty} r^{2n} |a_n|^2,$$

and thus $f \in H^2$ if and only if $\|f\|_2^2 = \sum_{n=0}^{\infty} |a_n|^2 < \infty$. It is also clear that the functions $f_r \in L^2(\mathbb{T})$ defined by $f_r(e^{i\omega}) = f(re^{i\omega})$ converge in the L^2 norm as $r \to 1$ to the function $\tilde{f}$ with Fourier coefficients $(a_n)_{n \geq 0}$, and hence a subsequence converges pointwise almost everywhere. Conversely, any function $\tilde{f} \in L^2(\mathbb{T})$ whose Fourier coefficients of negative index all vanish corresponds in an obvious way to a function f in H^2.

For the case $p = \infty$, we note that $H^\infty \subset H^2$, and thus a function $f \in H^\infty$ also corresponds to a boundary function $\tilde{f} \in L^2(\mathbb{T})$. Because a subsequence of (f_r) tends pointwise almost everywhere to $\tilde{f}$, we may conclude that $\|\tilde{f}\|_\infty \leq \|f\|_\infty$. However, one can obtain the extension from $\tilde{f}$ to f by integrating with the *Poisson kernel* K_r, namely,

$$f(re^{it}) = \frac{1}{2\pi} \int_0^{2\pi} K_r(t - \omega) \tilde{f}(e^{i\omega}) \, d\omega,$$

where

$$K_r(t - \omega) = \frac{1 - r^2}{1 + r^2 - 2r\cos(t - \omega)} = \operatorname{Re}\left(\frac{e^{i\omega} + re^{it}}{e^{i\omega} - re^{it}}\right). \qquad (1.3)$$

This implies that

$$\|f\|_\infty \leq \sup_{0 \leq r < 1} \|\tilde{f}\|_\infty \|K_r\|_1 = \|\tilde{f}\|_\infty,$$

since the Poisson kernel is a positive function and

$$\|K_r\|_1 = \frac{1}{2\pi} \int_0^{2\pi} K_r(t) \, dt = 1 \qquad (0 \leq r < 1).$$

We thus have $\|\tilde{f}\|_\infty = \|f\|_\infty$. $\square$

The Poisson kernel above provides a harmonic extension to the disc for any function in $L^1(\mathbb{T})$: the function is in the Hardy class if and only if this extension is actually an analytic function in $\mathbb{D}$.

We see that H^2 is a Hilbert space, being a closed subspace of the Hilbert space $L^2(\mathbb{T})$, and we shall use P_{H^2} to denote the orthogonal projection from $L^2(\mathbb{T})$ onto H^2, so that

$$P_{H^2} : \sum_{n=-\infty}^{\infty} a_n e^{in\omega} \mapsto \sum_{n=0}^{\infty} a_n e^{in\omega}.$$

There is a natural isometric isomorphism between the sequence space $\ell^2(\mathbb{Z})$ and the function space $L^2(\mathbb{T})$, given by associating the sequence $(a_n)_{n=-\infty}^{\infty}$ with the function whose Fourier series is $\sum_{n=-\infty}^{\infty} a_n e^{in\omega}$. Under this correspondence, the sequence space $\ell^2(\mathbb{Z}_+)$ maps to the Hardy space H^2, for we may regard $\ell^2(\mathbb{Z}_+)$ as embedding into $\ell^2(\mathbb{Z})$ as the subspace of all $(a_n)_{n=-\infty}^{\infty}$ with $a_n = 0$ for $n < 0$.

The space H^2 is also a *reproducing kernel Hilbert space* on the disc $\mathbb{D}$. What this means is that the evaluation functional $f \mapsto f(a)$ is bounded for each $a \in \mathbb{D}$. By the Riesz–Fréchet theorem given in Section 1.1, we can find a function $k_a \in H^2$, the *reproducing kernel*, such that

$$f(a) = \langle f, k_a \rangle = \frac{1}{2\pi} \int_0^{2\pi} f(e^{i\omega}) \overline{k_a(e^{i\omega})} \, d\omega,$$

and in fact in this case $k_a(z) = 1/(1 - \bar{a}z)$.

Another useful result is the following, which describes the boundary behaviour of an H^p function. We shall omit the proof, although in Chapter 3 we shall see that the final statement can be proved using the theory of shift-invariant subspaces (see Exercise 6 of Chapter 3).

Theorem 1.2.3 *Suppose that $f \in H^p$ for some $p \geq 1$, and that f is not identically zero. Then*

$$\frac{1}{2\pi} \int_0^{2\pi} \log |f(e^{i\omega})| \, d\omega > -\infty,$$

and hence $f(e^{i\omega}) \neq 0$ almost everywhere.

We now present the analogous results for Hardy spaces defined on the right half-plane $\mathbb{C}_+ = \{z \in \mathbb{C} : \operatorname{Re} z > 0\}$.

Definition 1.2.4 *For $1 \leq p < \infty$, the Hardy space $H^p(\mathbb{C}_+)$ of the right half-plane $\mathbb{C}_+$ may be defined as the set of all analytic functions $f : \mathbb{C}_+ \to \mathbb{C}$ such* *that*

$$\|f\|_p = \left(\sup_{x>0} \int_{-\infty}^{\infty} |f(x+iy)|^p \, dy\right)^{1/p} < \infty.$$

Likewise, the space $H^\infty(\mathbb{C}_+)$ consists of all analytic and bounded functions in $\mathbb{C}_+$, and the norm is given by

$$\|f\|_\infty = \sup_{z\in\mathbb{C}_+} |f(z)|.$$

Again these functions have boundary values $\tilde{f}(iy) = \lim_{x\to 0+} f(x+iy)$ almost everywhere, and the boundary function $\tilde{f}$ lies in $L^p(i\mathbb{R})$ and satisfies $\|\tilde{f}\|_{L^p} = \|f\|_{H^p}$. We may identify f and $\tilde{f}$, and thus $H^p(\mathbb{C}_+)$ can naturally be regarded as a closed subspace of $L^p(i\mathbb{R})$ and hence a Banach space.

As in the case of the disc, a Poisson kernel formula can be used to provide harmonic extensions to the right half-plane of functions lying in $L^p(i\mathbb{R})$ for some $1 \le p \le \infty$, and the H^p functions are those whose harmonic extensions are analytic. The formula in this case is

$$f(x+iy) = \frac{1}{\pi} \int_{-\infty}^{\infty} \hat{K}_x(y-t)\tilde{f}(it) \, dt, \qquad \text{for } x > 0,$$

where $\hat{K}_x$ is the *Poisson kernel for $\mathbb{C}_+$* and is given by

$$\hat{K}_x(y-t) = \frac{x}{x^2 + (y-t)^2}.$$

The *Laplace transform $L : L^2(0,\infty) \to H^2(\mathbb{C}_+)$* plays an important role here. It is given by

$$(Lg)(s) = \int_0^\infty e^{-st} g(t) \, dt \qquad (s \in \mathbb{C}_+)$$

and up to a constant factor gives an isometric isomorphism between the two spaces, since it is bijective and satisfies $\|Lg\|_{H^2} = \sqrt{2\pi}\|g\|_{L^2}$. What is even more remarkable is the content of the Paley–Wiener theorem, namely, that up to a change of variable we may use the bilateral Laplace transform (which is the Fourier transform with a change of variable) to decompose $L^2(i\mathbb{R})$ into the orthogonal sum of Hardy spaces on the left and right half-planes. Explicitly, if we now write

$$(Lg)(s) = \int_{-\infty}^{\infty} e^{-st} g(t) \, dt$$

for $s \in i\mathbb{R}$ and $g \in L^1(\mathbb{R})$, this extends by continuity to a linear isomorphism $L : L^2(\mathbb{R}) \to L^2(i\mathbb{R})$ with $\|Lg\|_{L^2} = \sqrt{2\pi}\|g\|_{L^2}$, and applying L to the decomposition

$$L^2(\mathbb{R}) = L^2(-\infty, 0) \oplus L^2(0, \infty)$$

gives

$$L^2(i\mathbb{R}) \;\; = \;\; H^2(\mathbb{C}_-) \oplus H^2(\mathbb{C}_+),$$

where $\mathbb{C}_-$ and $\mathbb{C}_+$ are, respectively, the left and right half-planes. We identify functions with their harmonic extensions, as before.

We note that $H^2(\mathbb{C}_+)$ is again a reproducing kernel Hilbert space, in that, for $s \in \mathbb{C}_+$ and $f \in H^2(\mathbb{C}_+)$, we have

$$f(s) = \langle f, k_s \rangle = \int_{-\infty}^{\infty} f(iy)\overline{k_s(iy)}dy,$$

where $k_s(z) = \dfrac{1}{2\pi(z + \bar{s})}$ is the reproducing kernel for $H^2(\mathbb{C}_+)$.

There is a natural isometric isomorphism between Hardy spaces on the disc and half-plane, which is induced by the self-inverse bijection $M : \mathbb{D} \to \mathbb{C}_+$, given by $M(z) = (1-z)/(1+z)$. The following relation is given in [57, 97], for example.

Theorem 1.2.5 *The mapping* $V : H^2(\mathbb{D}) \to H^2(\mathbb{C}_+)$, *defined by*

$$(Vf)(s) = \frac{1}{\sqrt{\pi}(1 + s)}f(M(s)),$$

is an isometric isomorphism.

1.3   Inner and outer functions

In this section we are concerned with the multiplicative structure of the Hardy spaces, in that we want to factorize a general Hardy class function as the product of two somewhat simpler functions, an inner factor and an outer factor. Here are their definitions (a simpler characterization of outer functions appears as a corollary of Beurling's theorem, in Corollary 3.1.4).

Definition 1.3.1 *An* inner function *is an* H^∞ *function that has unit modulus almost everywhere on* $\mathbb{T}$. *An* outer function *is a function* $f \in H^1$ *that can be written in the form*

$$f(z) = \alpha \exp\left(\frac{1}{2\pi} \int_0^{2\pi} \frac{e^{i\omega} + z}{e^{i\omega} - z} k(e^{i\omega}) \, d\omega \right) \qquad (z \in \mathbb{D}), \qquad (1.4)$$

where k *is a real-valued integrable function and* $|\alpha| = 1$.

If f is an outer function satisfying (1.4), then $\log|f(e^{it})| = k(e^{it})$ almost everywhere, since $\log|f(re^{it})| = \operatorname{Re}\log f(re^{it})$ gives the Poisson extension of k to the disc by virtue of (1.3). Clearly, an outer function can have no zeroes in the disc, since it is the exponential of something.

Example 1.3.2 *Examples of outer functions include all invertible functions in H^∞ (for example, polynomials whose zeroes all lie outside $\overline{\mathbb{D}}$). In fact it can be shown that all polynomials whose zeroes lie in $\mathbb{C}\setminus\mathbb{D}$ are outer functions, although they are not invertible in H^∞ if they have zeroes on the unit circle.*

Examples of inner functions include functions such as $z \mapsto \dfrac{z-a}{1-\overline{a}z}$, where $a \in \mathbb{D}$, and, more generally, Blaschke products (see Definition 1.3.4 below); these have zeroes in the disc, but there are also inner functions without zeroes, such as $\exp((z-1)/(z+1))$. This last function is just e^{-s}, where $s = (1-z)/(1+z)$; the mapping from z to s takes $\mathbb{D}$ to the right half-plane $\mathbb{C}_+$ and $\mathbb{T}\setminus\{-1\}$ to the imaginary axis $i\mathbb{R}$.

Once we have a complete characterization of inner functions and outer functions, we have a full description of all Hardy class functions, because of the following factorization theorem, which decomposes an arbitrary function into the product of inner and outer factors.

Theorem 1.3.3 (Inner–outer factorization) *Let f be a nonzero function in H^1. Then f has a factorization $f = \theta \cdot u$, where θ is inner and u is outer. This factorization is unique up to a constant of modulus 1.*

Sketch proof: Since $f \in H^1$, the function $\log|f|$ lies in $L^1(\mathbb{T})$, by virtue of Theorem 1.2.3. We can thus define the outer factor u corresponding to f by the formula

$$u(z) = \exp\left(\frac{1}{2\pi}\int_0^{2\pi} \frac{e^{i\omega}+z}{e^{i\omega}-z}\log|f(e^{i\omega})|\,d\omega\right), \qquad (z \in \mathbb{D}),$$

after which $\theta = f/u$ is analytic in $\mathbb{D}$ with boundary values of modulus 1 almost everywhere, and thus θ is inner. The uniqueness of the decomposition follows on observing that a unimodular outer function is necessarily constant. $\qquad\square$

We are going to see a complete description of the class of inner functions, and we begin with those that have zeroes in $\mathbb{D}$.

Definition 1.3.4 *A finite Blaschke product is a function of the form*

$$B(z) = \alpha \prod_{j=1}^{n} \frac{z-z_j}{1-\overline{z}_j z},$$

where $|\alpha| = 1$ and $|z_j| < 1$ for $j = 1,\ldots,n$.

It is easy to verify that B is analytic in $\mathbb{D}$ and continuous in $\overline{\mathbb{D}}$, that B is inner, and that B has zeroes at $z_1, \ldots, z_n$ only and poles at $1/\overline{z}_1, \ldots, 1/\overline{z}_n$ only.

Next we want to break the inner part into two factors, an inner function with zeroes (which will be an infinite Blaschke product) and an inner function without zeroes (a so-called *singular inner function*). To do this we need to understand the properties of the zero set of a function in H^p.

Theorem 1.3.5 (G. Szegő) *Let $f \in H^1$ be such that f is not identically zero. Then the zeroes (z_n) of f are countable in number and satisfy the Blaschke condition*

$$\sum_{1}^{\infty} (1 - |z_n|) < \infty. \tag{1.5}$$

Proof: By considering $z \mapsto f(z)/z^j$, if necessary, we may suppose without loss of generality that $f(0) \neq 0$. Now take $0 < r < 1$ and let $z_1, \ldots, z_m$ be the zeroes of f in $\{z \in \mathbb{C} : |z| < r\}$, choosing r so that there are none on $\{|z| = r\}$. Write $g(z) = f(rz)/B(z)$, where B is a Blaschke product with zeroes $z_1/r, \ldots, z_m/r$. Since $\log g$ is harmonic, we have the identity

$$\log g(0) = \frac{1}{2\pi} \int_0^{2\pi} \log g(e^{i\omega}) \, d\omega,$$

which, on taking real parts, reduces to

$$\log |f(0)| + \sum_{|z_n| < r} \log(r/|z_n|) = \frac{1}{2\pi} \int_0^{2\pi} \log |f(re^{i\omega})| \, d\omega \leq \log \|f\|_1,$$

by Jensen's inequality $\int \phi(f(x)) \, dx \leq \phi \int f(x) \, dx$ holding for concave functions ϕ – in this case $\phi(y) = \log(y)$.

Letting $r \to 1$, we see that $\sum_n \log(1/|z_n|) < \infty$, which, by the comparison test, is easily seen to be equivalent to $\sum(1 - |z_n|) < \infty$.   □

For a Hardy class function f we can construct a Blaschke product B whose zeroes are precisely the zeroes of f; then f/B has no zeroes at all and can be analysed further.

Theorem 1.3.6 *Let $f \in H^1$. Then the infinite Blaschke product*

$$B(z) = z^m \prod_{|z_n| \neq 0} \frac{\overline{z}_n}{|z_n|} \frac{z_n - z}{1 - \overline{z}_n z},$$

where (z_n) are the zeroes of f, of which m are at 0, converges uniformly on compact sets to an H^∞ function, the only zeroes of which are the (z_n), with the correct multiplicities. Moreover, $|B(z)| \leq 1$ and $|B(e^{i\omega})| = 1$ almost everywhere.

Proof: We write b_n for the factor corresponding to z_n in the infinite product formula for B. Note that

$$|1 - b_n(z)| = \frac{(1 - |z_n|)\,|\bar{z}_n z + |z_n||}{|z_n|\,|1 - \bar{z}_n z|} \leq (1 - |z_n|)\frac{1 + |z|}{1 - |z|},$$

which, combined with the Blaschke condition (1.5), guarantees local uniform convergence of the infinite product to a function B with the required zeroes.

It now remains to verify that B is also inner. Clearly $|B(e^{i\omega})| \leq 1$ for all $\omega \in [0, 2\pi]$; further, letting $B_n = B/(b_1 \ldots b_n)$, which is also a Blaschke product, we have

$$|B_n(0)| \leq \frac{1}{2\pi} \int_0^{2\pi} |B_n(e^{i\omega})|\, d\omega = \frac{1}{2\pi} \int_0^{2\pi} |B(e^{i\omega})|\, d\omega,$$

and letting $n \to \infty$ we see that

$$\frac{1}{2\pi} \int_0^{2\pi} |B(e^{i\omega})|\, d\omega = 1,$$

so that $|B(e^{i\omega})| = 1$ almost everywhere. □

Removing the Blaschke factor leaves a function without zeroes and does not change the H^p norm, since the boundary values of an inner function are unimodular almost everywhere. We thus obtain the following factorization result.

Corollary 1.3.7 (F. Riesz) *Let $f \in H^p$, $f \not\equiv 0$, and let $B(z)$ be the (possibly infinite) Blaschke product formed using the zeroes (z_n) of f. Then $f(z) = g(z)B(z)$ for some $g \in H^p$ with $\|f\|_p = \|g\|_p$. It follows that any nonzero function $f \in H^1$ can be written as $f = B \cdot S \cdot u$, where B is a Blaschke product, S is a singular inner function (one without zeroes), and u is an outer function. This factorization is unique up to constants of modulus 1.*

The next result explains why an inner function without zeroes is called a singular inner function. Formula (1.6) below is very similar to (1.4), except that the integral is now taken with respect to a singular measure, that is, one supported on a set of Lebesgue measure 0, rather than $k(\omega)\, d\omega$.

Theorem 1.3.8 *Let g be an inner function without zeroes. Then there is a unique positive measure μ, singular with respect to Lebesgue measure, and a constant α of modulus 1, such that*

$$g(z) = \alpha \exp\left(-\int_0^{2\pi} \frac{e^{i\omega} + z}{e^{i\omega} - z}\, d\mu(\omega)\right), \qquad (z \in \mathbb{D}). \tag{1.6}$$

Sketch proof: Choose a branch of $h = -\log g$ in $\mathbb{D}$, since g has no zeroes: we see that $\operatorname{Re} h = -\log|g| \geq 0$ in $\mathbb{D}$. We now use the theorem of Herglotz, which asserts that any non-negative harmonic function ϕ in $\mathbb{D}$ is the Poisson integral of a positive measure μ on $\mathbb{T}$; indeed, we can identity $d\mu(\omega)$ as the weak-* limit point as $r \to 1$ of a sequence of measures $\frac{1}{2\pi}\phi_r \, d\omega$, where $\phi_r(e^{i\omega}) = \phi(re^{i\omega})$. It follows that $g = e^{-h}$, where

$$h(z) = i\beta + \int_0^{2\pi} \frac{e^{i\omega} + z}{e^{i\omega} - z} \, d\mu(\omega), \qquad (z \in \mathbb{D}),$$

and β is a real constant. The fact that μ is a singular measure follows from the fact that g is inner, and thus the radial limits of $\operatorname{Re} h$, which (up to a constant) equal $\frac{d\mu}{d\omega}$, are zero almost everywhere. $\square$

A similar inner–outer factorization holds for the Hardy spaces of the right half-plane. The Blaschke condition on the zeroes (s_n) of an $H^p(\mathbb{C}_+)$ function is now the following:

$$\sum_n \frac{\operatorname{Re} s_n}{1 + |s_n|^2} < \infty,$$

and the corresponding Blaschke product has the form

$$B(s) = \left(\frac{s-1}{s+1}\right)^m \prod_{s_n \neq 1} \frac{|1 - s_n|^2}{1 - s_n^2} \frac{s - s_n}{s + \overline{s}_n}.$$

A non-zero inner function is again represented by a measure supported on the boundary, but in this case the measure can have mass "at infinity", which manifests itself as an extra factor $e^{-\lambda s}$. The general form of a singular inner function is thus:

$$g(s) = e^{-\lambda s} \exp\left(-\int_{i\mathbb{R}} \frac{ys + i}{y + is} \, d\mu(y)\right),$$

where μ is now a measure on $\mathbb{R}$. For example, $e^{-1/s}$ is an inner function, corresponding to a point mass at the origin.

Outer functions in $H^p(\mathbb{C}_+)$ have the form

$$f(s) = \alpha \exp\left(\frac{1}{\pi} \int_{-\infty}^{\infty} \frac{ys + i}{y + is} k(iy) \frac{dy}{1 + y^2}\right), \tag{1.7}$$

where $|\alpha| = 1$, the function k is real valued, and

$$\int_{-\infty}^{\infty} |k(iy)| \frac{dy}{1 + y^2} \qquad \text{converges.}$$

For such a function f we have $\log|f(iy)| = k(iy)$ almost everywhere.

1.4 Vector-valued Hardy spaces

It is possible to define Hardy spaces of functions taking values in arbitrary Banach spaces, but this introduces certain complications when we come to look at integration. However, in the case of finite-dimensional Banach spaces, there is no such difficulty, and we shall restrict ourselves to these. From a linear systems point of view, this is the most interesting case, since this is a suitable framework for modelling systems with a finite number of inputs and outputs.

A function F with values in an n-dimensional complex vector space $\mathcal{V}$ clearly has the form $F(z) = f_1(z)v_1 + f_2(z)v_2 + \ldots + f_n(z)v_n$, where $f_1, \ldots, f_n$ are ordinary complex-valued functions and $v_1, \ldots, v_n$ is a basis for $\mathcal{V}$. We then say that F is analytic if and only if each of $f_1, \ldots, f_n$ is analytic. The definition is easily seen to be independent of the choice of basis.

In fact, there are two vector-valued Hardy spaces that principally interest us here. For the first, we consider functions with values in $\mathbb{C}^n$, which we provide with the standard Euclidean norm.

Definition 1.4.1 *The space $H^2(\mathbb{D}, \mathbb{C}^n)$ consists of all analytic functions $F : \mathbb{D} \to \mathbb{C}^n$, such that the norm given by*

$$\|F\| = \left(\sup_{0 < r < 1} \frac{1}{2\pi} \int_0^{2\pi} \|f(re^{i\omega})\|^2 \, d\omega \right)^{1/2}$$

is finite.

It is easily seen that this embeds in a linear and isometric fashion into $L^2(\mathbb{T}, \mathbb{C}^n)$, much as in the scalar case. Indeed, if $F = (f_1, \ldots, f_n)$, then

$$\|F\|^2 = \|f_1\|^2 + \ldots + \|f_n\|^2.$$

The other important example is a matrix-valued version of H^∞. Recall that if m and n are positive integers, then we may identify $\mathcal{L}(\mathbb{C}^m, \mathbb{C}^n)$ with the vector space of all $n \times m$ matrices. To do this, we associate the operator T with the matrix (a_{jk}) by the formula $Te_j = \sum_{k=1}^n a_{kj} e'_k$, where $e_1, \ldots, e_m$ and $e'_1, \ldots, e'_n$ are the usual orthonormal bases of $\mathbb{C}^m$ and $\mathbb{C}^n$, respectively. The operator norm of T is given by

$$\|T\| = \max\{\sigma_1, \ldots, \sigma_m\},$$

where $\sigma_1, \ldots, \sigma_m$ are the singular values of T, as in Section 1.1, that is, $\sigma_1^2, \ldots, \sigma_m^2$ are the eigenvalues of the non-negative self-adjoint matrix T^*T. Sometimes we write $\sigma_1(T)$ for $\|T\|$ to avoid a proliferation of norm symbols, as in the following definition.

Definition 1.4.2 *The space $H^\infty(\mathbb{D}, \mathcal{L}(\mathbb{C}^m, \mathbb{C}^n))$ consists of all bounded analytic functions $F : \mathbb{D} \to \mathcal{L}(\mathbb{C}^m, \mathbb{C}^n)$, with the norm given by*

$$\|F\| = \sup_{z \in \mathbb{D}} \sigma_1(F(z)).$$

Again, we may embed $H^\infty(\mathcal{L}(\mathbb{C}^m, \mathbb{C}^n))$ linearly and isometrically into the space $L^\infty(\mathbb{T}, \mathcal{L}(\mathbb{C}^m, \mathbb{C}^n))$.

Similarly, vector-valued Hardy spaces on the right half-plane may be defined, such as $H^2(\mathbb{C}_+, \mathbb{C}^n)$ and $H^\infty(\mathbb{C}_+, \mathcal{L}(\mathbb{C}^m, \mathbb{C}^n))$. We leave the reader to write down the analogous definitions when required.

Notes

The elementary theory of Hilbert spaces can be found in [9, 48, 117, 118, 149, 146] and in many other places.

The basic ideas in the theory of Hardy spaces are also standard and can be found in various forms in [2, 24, 26, 39, 57, 97, 100, 116], for example.

The inner–outer factorization should be compared with related decompositions occurring elsewhere in analysis: for example, any bounded operator A has a *polar decomposition* $A = V|A|$, where $|A| = (A^*A)^{1/2}$ is a positive operator and V is a *partial isometry* (an operator that is an isometry from $(\ker A)^\perp$ onto $(\ker A^*)^\perp$ and zero on $\ker A$).

Exercises

1. Prove that if T is a linear operator that is continuous at 0, then there is a number $\delta > 0$ such that $\|Tx\| \leq 1$ whenever $\|x\| \leq \delta$, and deduce that T is bounded with $\|T\| \leq 1/\delta$.

2. Show that if S and T are elements of $\mathcal{L}(\mathcal{X})$, then $\|ST\| \leq \|S\| \|T\|$.

3. The *Volterra operator* $V : L^2(0,1) \to L^2(0,1)$ is defined by

$$(Vf)(x) = \int_0^x f(t)\,dt \qquad (f \in L^2(0,1)).$$

Use the Cauchy–Schwarz inequality to show that $|(Vf)(x)| \leq \sqrt{x}\|f\|_2$. Deduce that $\|V\| \leq 1/\sqrt{2}$.

4. Let $R : \ell^2 \to \ell^2$ denote the right shift, that is,

$$R(a_1, a_2, \ldots) = (0, a_1, a_2, \ldots).$$

Prove that $\|R\| = 1$, so that $\sigma(R) \subseteq \overline{\mathbb{D}}$. Verify that $\sigma_p(R) = \emptyset$, that is, R has no eigenvalues. Show, however, that $\sigma(R) = \overline{\mathbb{D}}$.

5. Let $(\lambda_n)_{n=0}^\infty$ be a fixed bounded sequence of complex numbers, and define an operator on $\ell^2(\mathbb{Z}_+)$ by $T((x_n)) = ((y_n))$, where $y_n = \lambda_n x_n$ for each n. Verify that T is a bounded operator and $\|T\| = \|(\lambda_n)\|_\infty$. Let $\Lambda = \{\lambda_1, \lambda_2, \ldots\}$. Prove that each λ_k is an eigenvalue of T, and hence is in $\sigma(T)$, and that if $\lambda \notin \overline{\Lambda}$, then the inverse of $T - \lambda I$ exists and is bounded. Deduce that $\sigma(T) = \overline{\Lambda}$.

6. Find the adjoint of the rank-one operator T, defined by $Tx = \langle x, y \rangle z$, $(x \in \mathcal{H})$, where y and z are fixed elements of a Hilbert space $\mathcal{H}$.

7. Let $U \in \mathcal{L}(\mathcal{H})$ be a unitary operator. Show that (Ue_n) is an orthonormal basis of $\mathcal{H}$ whenever (e_n) is.

8. Let $f \in C[a, b]$, the space of continuous functions on an interval $[a, b]$, and let M_f be the multiplication operator on $L^2(a, b)$, given by $(M_f g)(t) = f(t)g(t)$, for $g \in L^2(a, b)$. Find a function $\tilde{f} \in C[a, b]$ such that $M_f^* = M_{\tilde{f}}$. Show that M_f is always a normal operator. When is it Hermitian? When is it unitary?

9. Let $T \in \mathcal{L}(\mathcal{H})$. Show that $\sigma(T^*) = \{\overline{\lambda} : \lambda \in \sigma(T)\}$.

10. Let R be as in Exercise 4. Prove that $R^* : \ell^2 \to \ell^2$ is given by the left shift $R^*(b_1, b_2, \ldots) = (b_2, b_3, \ldots)$. Show that $\sigma_p(R^*) = \mathbb{D}$. Deduce that $\sigma(R^*) = \overline{\mathbb{D}}$. Finally, show that R is not a normal operator.

11. Show, using the Cauchy–Schwarz inequality, that $H^\infty \subseteq H^2 \subseteq H^1$, and give examples to show that the inclusions are strict.

12. Suppose that $f \in L^2(\mathbb{T})$ has the Fourier series $\sum_{n=-\infty}^{\infty} a_n e^{in\omega}$. Show that its harmonic extension to the disc is given by $f(re^{it}) = \sum_{n=-\infty}^{\infty} a_n r^{|n|} e^{int}$.

13. Let $f(z) = \log((1-z)/(1+z))$ for $z \in \mathbb{D}$. Is $\operatorname{Re} f$ bounded in $\mathbb{D}$? Is $\operatorname{Im} f$ bounded in $\mathbb{D}$? Show that f lies in H^2 but not in H^∞.

14. For $a \in \mathbb{D}$, let $k_a : z \mapsto 1/(1 - \bar{a}z)$ be the reproducing kernel. Verify that $\langle f, k_a \rangle = f(a)$ for all $f \in H^2$. What is $\|k_a\|_2$?

15. For $\lambda \in \mathbb{C}_+$, calculate the Laplace transform of the function $f(t) = e^{-\lambda t}$ in $L^2(0, \infty)$ and verify that $\|Lf\|_{H^2(\mathbb{C}_+)} = \sqrt{2\pi}\|f\|_{L^2(0,\infty)}$.

16. Verify that the function $s \mapsto e^{-s}/(s-1)$ is in $L^2(i\mathbb{R})$. Calculate its orthogonal projection onto $H^2(\mathbb{C}_+)$.

17. Find necessary and sufficient conditions on a rational function $f(z) = c\prod_{j=1}^{m}(z - a_j)/\prod_{k=1}^{n}(z - b_k)$ for it to be in $H^\infty(\mathbb{D})$ and (a) inner and (b) outer.

18. For the inner function $\theta(s) = e^{-s}$ in $H^\infty(\mathbb{C}_+)$, show that $\theta H^2(\mathbb{C}_+)$ consists of all functions that are Laplace transforms of functions in $L^2(1, \infty)$, and that $H^2(\mathbb{C}_+) \ominus \theta H^2(\mathbb{C}_+)$ corresponds to $L^2(0, 1)$ in the same way.

19. Fill in the details of the sketched proofs in Section 1.3.

Chapter 2

Closed Operators

The main theme of this chapter is the study of unbounded operators by means of their graphs, that is, sets of pairs (x, Tx). Unbounded operators occur in many applications: for example, the integral operator,

$$y(t) = \int_0^t u(t)\, dt,$$

which models a component of so many physical systems, sometimes by means of the first-order differential equation

$$\frac{dy}{dt} = u(t), \qquad y(0) = 0,$$

is unbounded on $L^2(0, \infty)$ and can be successfully studied by these techniques.

More glamorous examples of unbounded operators occur in the theory of semigroups (for example, the heat semigroup), and we study these in greater detail. The gap metric is introduced here, and it will play a fundamental role in the control-theoretic ideas of Chapter 4.

2.1 The graph of an operator

In order to treat unbounded operators using methods of analysis, rather than purely algebraic techniques, we now introduce the *graph* of an operator. Note that the term is being used in the elementary sense of *the graph of the function* $y = 4x$, rather than as a branch of graph theory [8].

Definition 2.1.1 *Let* $T : \mathcal{X} \to \mathcal{Y}$ *be a mapping between sets. Then its* graph *is the subset* $\mathcal{G}(T) \subseteq \mathcal{X} \times \mathcal{Y}$ *given by*

$$\mathcal{G}(T) = \{(x, Tx) : x \in \mathcal{X}\}.$$

If $\mathcal{X}$ and $\mathcal{Y}$ are vector spaces and T is linear, then it is straightforward to verify that $\mathcal{G}(T)$ is a linear subspace of the vector space $\mathcal{X} \times \mathcal{Y}$ (see the exercises).

Let us now look at the case when $\mathcal{X}$ and $\mathcal{Y}$ are normed spaces and T is linear. Then $\mathcal{X} \times \mathcal{Y}$ is naturally a normed space, with the topology given by a number of possible norms; for example, for $1 \le p < \infty$ we may define

$$\|(x,y)\|_p = (\|x\|^p + \|y\|^p)^{1/p} \tag{2.1}$$

(the case $p = 2$ will be convenient when $\mathcal{X}$ and $\mathcal{Y}$ are inner-product spaces, because then $\mathcal{X} \times \mathcal{Y}$ is also an inner-product space in a natural way). Another possible norm is given by

$$\|(x,y)\|_\infty = \max\left(\|x\|, \|y\|\right).$$

An operator is then said to be *closed* if its graph is closed. This is equivalent to the condition that, whenever (x_n) is a sequence in $\mathcal{X}$ and x, y are vectors in $\mathcal{X}$, $\mathcal{Y}$, respectively, such that $x_n \to x$ and $Tx_n \to y$, then we have $y = Tx$. Clearly any continuous (i.e., bounded) operator is closed, and the following closed graph theorem of Banach gives a partial converse in the case when $\mathcal{X}$ and $\mathcal{Y}$ are Banach spaces (i.e., complete normed spaces). A proof can be found in many books, e.g., [116, 146].

Theorem 2.1.2 (Closed graph theorem) *Let $T : \mathcal{X} \to \mathcal{Y}$ be a linear mapping between Banach spaces such that $\mathcal{G}(T)$ is a closed subspace of $\mathcal{X} \times \mathcal{Y}$. Then T is bounded.*

When we come to look at unbounded operators, there is one complication that needs to be tackled. Although it is possible to show the existence of everywhere-defined unbounded operators between Banach spaces (see the exercises), the proof is not constructive, and the operators that one encounters in applications are normally defined only on an incomplete subspace of a Banach space.

Let $\mathcal{X}$ and $\mathcal{Y}$ be normed spaces; if T is an operator defined on a subspace of $\mathcal{X}$, then we call that subspace the *domain of T* and denote it by $\mathcal{D}(T)$. However, it is still common to talk about an operator between $\mathcal{X}$ and $\mathcal{Y}$, even though it is not defined on the whole of $\mathcal{X}$.

It is easy to see that a linear subspace $\mathcal{S} \subseteq \mathcal{X} \times \mathcal{Y}$ is the graph of an operator T (defined on some subspace $\mathcal{D}(T) \subseteq \mathcal{X}$) if and only if, whenever we have $(x, z_1) \in \mathcal{S}$ and $(x, z_2) \in \mathcal{S}$, we necessarily have $z_1 = z_2$ or, equivalently, if the only vector $(0, y)$ lying in $\mathcal{S}$ is $(0, 0)$.

Example 2.1.3 Let $\mathcal{H}$ be a Hilbert space with orthonormal basis $(e_n)_{n=1}^{\infty}$ and define an operator T by $T : \sum_{n=1}^{\infty} a_n e_n \mapsto \sum_{n=1}^{\infty} n a_n e_n$. Then T is unbounded, since $T e_n = n e_n$, and T is not defined on the whole of $\mathcal{H}$, indeed

$$\mathcal{D}(T) = \left\{ \sum_{n=1}^{\infty} a_n e_n \in \mathcal{H} : \sum_{n=1}^{\infty} n a_n e_n \in \mathcal{H} \right\},$$

so that $\sum_{n=1}^{\infty} a_n e_n \in \mathcal{D}(T)$ if and only if $\sum_{n=1}^{\infty} n^2 |a_n|^2 < \infty$. Thus $\mathcal{D}(T)$ is a dense subspace of $\mathcal{H}$ (since it includes all finite linear combinations of the vectors e_n) but not the whole of $\mathcal{H}$.

Now, even though $\mathcal{D}(T)$ is not closed in Example 2.1.3, it turns out that the graph of T is closed. This subspace is given by

$$\mathcal{G}(T) = \{(x, Tx) : x \in \mathcal{D}(T)\},$$

and it is this definition that will apply in general. To see that the graph is closed, suppose that $(x^{(k)})_{k=1}^{\infty}$ and x lie in $\mathcal{D}(T)$, with $x^{(k)} \to x$ and $T x^{(k)} \to y$, for some vector $y \in \mathcal{H}$. (This superscript notation for sequences is ugly but is convenient here.) Let $x = \sum_{n=1}^{\infty} a_n e_n$ and $y = \sum_{n=1}^{\infty} b_n e_n$. Then for each n the nth coordinate of $x^{(k)}$ tends to a_n and the nth coordinate of $T x^{(k)}$ tends to b_n. Hence $b_n = n a_n$ and thus $y = Tx$.

If $T : \mathcal{D}(T) \to \mathcal{Y}$ is an operator with a closed graph $\mathcal{G}(T) \subseteq \mathcal{X} \times \mathcal{Y}$, where $\mathcal{D}(T) \subseteq \mathcal{X}$, and $\mathcal{X}$ and $\mathcal{Y}$ are Banach spaces, then $\mathcal{D}(T)$ itself becomes a Banach space with the *graph norm*,

$$\|x\|_g = (\|x\|^2 + \|Tx\|^2)^{1/2}$$

and a Hilbert space if $\mathcal{X}$ and $\mathcal{Y}$ are Hilbert spaces. This follows because then $\mathcal{D}(T)$ is isometrically isomorphic to $\mathcal{G}(T)$ under the mapping $x \mapsto (x, Tx)$, if we take $p = 2$ in (2.1). Note that $\mathcal{G}(T)$ is a Banach space because it is a closed subspace of the Banach space $\mathcal{X} \times \mathcal{Y}$.

It sometimes happens that the graph of an operator T is not closed, but that its closure is still the graph of an operator. In this case we say that T is *closable* and define its *closure*, $\overline{T}$, to be the operator such that $\mathcal{G}(\overline{T}) = \overline{\mathcal{G}(T)}$. The following result tells us when an operator is closable.

Proposition 2.1.4 Let T be an operator with domain $\mathcal{D}(T) \subseteq \mathcal{X}$ and range $\mathcal{Y}$. Then T is closable if and only if, whenever (x_n) is a sequence in $\mathcal{D}(T)$ and y a vector in $\mathcal{Y}$ with $x_n \to 0$ and $Tx_n \to y$, we have $y = 0$.

Proof: Clearly the condition is necessary, since if $(0, y)$ lies in the closure of the graph of T, and if $\overline{\mathcal{G}(T)}$ is itself a graph, then $y = 0$. Conversely, if the condition is satisfied, then $\overline{\mathcal{G}(T)}$ is a linear subspace, since the closure of a linear subspace is always a linear subspace, and it satisfies the necessary and sufficient condition given above for a subspace to be the graph of an operator, namely, that it contains no vector $(0, y)$ with $y \neq 0$. $\square$

In particular, any bounded operator is closable, since it is continuous. Let us see an example of a non-closable operator.

Example 2.1.5 *In a Hilbert space $\mathcal{H}$ with orthonormal basis $(e_n)_{n=1}^{\infty}$, the vectors (e_n), together with an extra vector $v = \sum_{k=1}^{\infty} e_k/k$, form a linearly independent set (i.e., no finite linear combination vanishes). Let $\mathcal{S}$ denote its linear span, and define $T : \mathcal{S} \to \mathcal{H}$ by $Te_n = 0$ for all n and $Tv = v$, extending to $\mathcal{S}$ by linearity. Now the vectors $u_n = v - \sum_{k=1}^{n} e_k/k$ satisfy $u_n \to 0$ but $Tu_n = v$ for all n, and hence T is not closable.*

We discussed earlier the adjoint of a bounded operator on a Hilbert space, as defined in (1.1). It will now be useful to see the extent to which we can make such a definition for unbounded operators, which may not be defined on the whole space.

Definition 2.1.6 *Let $\mathcal{H}$ and $\mathcal{K}$ be Hilbert spaces, and let $T : \mathcal{D}(T) \to \mathcal{K}$ be a linear operator with domain $\mathcal{D}(T) \subseteq \mathcal{H}$. Then $T' : \mathcal{D}(T') \to \mathcal{H}$ is said to be an adjoint to T if $\mathcal{D}(T') \subseteq \mathcal{K}$ and*

$$\langle Th, k \rangle = \langle h, T'k \rangle \qquad \text{for all } h \in \mathcal{D}(T) \text{ and } k \in \mathcal{D}(T').$$

As it stands, there will be many adjoints; indeed, the definition is satisfied if we define $\mathcal{D}(T') = \{0\}$ and $T'(0) = 0$. However, suppose now that $\mathcal{D}(T)$ is dense and let us define T' as follows. Take $\mathcal{D}(T')$ to be the set of all k for which there actually exists an $h_0 \in \mathcal{H}$ with $\langle Th, k \rangle = \langle h, h_0 \rangle$ for all $h \in \mathcal{D}(T)$. Such an h_0 is unique, since $\mathcal{D}(T)$ is dense; thus $\langle h, h_0 \rangle$ is completely determined once we know its values on $\mathcal{D}(T)$.

Accordingly, we define $T'k = h_0$; it is easily verified that this makes T' into a linear operator adjoint to T, and that it is maximal, in the sense that all other operators adjoint to T' are restrictions of T' to a subdomain. We now write T^* for T' and call it *the adjoint* of T. It is not hard to verify that this definition produces the usual adjoint, in the case that T is bounded and defined on the whole of $\mathcal{H}$.

Remark 2.1.7 Even if T is not *a priori* closed, it is closable as soon as it has a densely-defined adjoint T'. For if $u_n \to 0$ and $Tu_n \to y$, then

$$\langle y, w \rangle = \lim_{n \to \infty} \langle Tu_n, w \rangle = \lim_{n \to \infty} \langle u_n, T'w \rangle = 0$$

for all $w \in \mathcal{D}(T')$ and so $y = 0$.

A further relevance of the adjoint to our discussion of graphs is explained in the next result. Note that $\mathcal{G}(T) \subseteq \mathcal{H} \times \mathcal{K}$ and $\mathcal{G}(T^*) \subseteq \mathcal{K} \times \mathcal{H}$; for convenience we write $\mathcal{G}'(T^*)$ for the *reversed graph* of T^*, that is,

$$\mathcal{G}'(T^*) = \{(h, k) \in \mathcal{H} \times \mathcal{K} : k \in \mathcal{D}(T^*), \, h = T^*k\}.$$

Theorem 2.1.8 *Let $\mathcal{H}$ and $\mathcal{K}$ be Hilbert spaces, and let $T : \mathcal{D}(T) \to \mathcal{K}$ be a closed linear operator with dense domain $\mathcal{D}(T) \subseteq \mathcal{H}$. Then $\mathcal{G}'(-T^*)$ is the orthogonal complement of $\mathcal{G}(T)$ in $\mathcal{H} \times \mathcal{K}$.*

Proof: We note that a vector (h_1, k_1) is orthogonal to the whole of $\mathcal{G}(T)$ if and only if $\langle (h, Th), (h_1, k_1) \rangle = 0$ for all $h \in \mathcal{D}(T)$. This gives the condition

$$\langle h, h_1 \rangle + \langle Th, k_1 \rangle = 0. \tag{2.2}$$

Now if $h_1 = -T^*k_1$, then we certainly have $\langle h, h_1 \rangle = -\langle Th, k_1 \rangle$ for all $h \in \mathcal{D}(T)$, and so (2.2) holds. On the other hand, if (2.2) does hold, then k_1 is in the domain of T^*, because we can define $T'k_1 = -h_1$ to satisfy the criterion in Definition 2.1.6. Thus $T^*k_1 = -h_1$. $\qquad\square$

Noting that orthogonal complements are themselves closed subspaces, the duality between T and T^* can clearly be expressed in a more symmetrical form, as follows.

Corollary 2.1.9 *Let $\mathcal{H}$ and $\mathcal{K}$ be Hilbert spaces, and let $T : \mathcal{D}(T) \to \mathcal{K}$ be a closed densely-defined operator with domain $\mathcal{D}(T) \subseteq \mathcal{H}$. Then T^* is a closed densely-defined operator with domain $\mathcal{D}(T^*) \subseteq \mathcal{K}$.*

Proof: It is clear from Theorem 2.1.8 that T^* has a closed graph. Moreover, T^* is densely defined, as otherwise there would exist a vector $k \neq 0$ with $k \perp \mathcal{D}(T^*)$; but then $(0, k) \in \mathcal{G}'(-T^*)^\perp$, which is $\mathcal{G}(T)$. This is impossible, since $\mathcal{G}(T)$ is the graph of an operator. $\qquad\square$

We leave the reader to verify (by looking at graphs) that the identity $(T^*)^* = T$ still holds in this case.

2.2   Semigroups

A very important class of closed unbounded operators arises in the theory of
semigroups of operators, which itself has strong links with the theory of linear
systems. Accordingly, we discuss some of the basic results here.

Definition 2.2.1 *Let $\mathcal{X}$ be a Banach space. Then a strongly continuous semi-
group(or C_0 semigroup) $(T(t))$ is a collection of bounded operators $\{T(t) : t \in$
$\mathbb{R}, \, t \geq 0\}$, satisfying the following conditions:*

1. *$T(0) = I$, the identity operator on $\mathcal{X}$.*

2. *$T(s)T(t) = T(s + t)$ for every $s, \, t \geq 0$.*

3. *The mapping from $\mathbb{R}_+$ into $\mathcal{X}$ defined by $t \mapsto T(t)x$ is continuous for every
   $x \in \mathcal{X}$.*

One important and easy example is obtained by defining $T(t) = e^{At}$, where
A is a fixed bounded operator on $\mathcal{X}$. In this case the mapping $t \mapsto T(t)$ is even
continuous in the norm topology (see the exercises). The following useful result
shows that the operators $T(t)$ cannot grow in norm faster than exponentially.

Lemma 2.2.2 *Let $(T(t))$ be a C_0 semigroup. Then there exist constants M,
$\alpha > 0$ such that $\|T(t)\| \leq Me^{\alpha t}$ for all $t \geq 0$.*

Proof: Consider first the operators $T(t)$ for $0 \leq t \leq 1$. These are uniformly
bounded in norm, since if (t_k) is any sequence in $[0, 1]$, then it has a convergent subsequence, $(t_{k(l)})$ say, converging to a point $u \in [0, 1]$. Then $(T(t_{k(l)})x)$
converges to $T(u)x$ for each $x \in \mathcal{X}$. By the Banach–Steinhaus (uniform boundedness) theorem [9, 117] the subsequence $(T(t_{k(l)}))$ is uniformly bounded in norm,
and this implies that there is a constant $K > 0$ such that $\|T(t)\| \leq K$ for
$0 \leq t \leq 1$. But now, given any real number $t \geq 0$, we write $t = n + r$, where n is
a non-negative integer and $0 \leq r < 1$. Then

$$\|T(t)\| \leq \|T(1)^n\| \, \|T(r)\| \leq K^{n+1} \leq Me^{\alpha t} \qquad \text{with} \quad M = K \quad \text{and} \quad e^\alpha = K.$$

$\square$

We shall now see that for any C_0 semigroup, even one not given *a priori* as
$T(t) = e^{At}$, an analogous operator A exists, although it will not in general be
bounded.

Definition 2.2.3 *Let $(T(t))$ be a C_0 semigroup on a Banach space $\mathcal{X}$. Then its
infinitesimal generator is the linear operator $A : \mathcal{D}(A) \to \mathcal{X}$ defined by*

$$Ax = \lim_{h \to 0+} \frac{T(h)x - x}{h},$$

with domain $\mathcal{D}(A) \subseteq \mathcal{X}$ given by

$$\mathcal{D}(A) = \{x \in \mathcal{X} : \lim_{h \to 0+} \frac{T(h)x - x}{h} \text{ exists}\}.$$

It is not hard to verify that $\mathcal{D}(A)$ is a linear subspace and that A is linear (see the exercises). The domain of A is also invariant under each $T(t)$, as the next result shows.

Proposition 2.2.4 *Let A be the infinitesimal generator of a C_0 semigroup $(T(t))$ defined on $\mathcal{X}$, and let $\mathcal{D}(A)$ denote its domain. Then for $z \in \mathcal{D}(A)$ one has $T(t)z \in \mathcal{D}(A)$, and $AT(t)z = T(t)Az$ for all $t \geq 0$.*

Proof: Checking the condition in Definition 2.2.3 gives, for $h > 0$,

$$\frac{T(h)T(t)z - T(t)z}{h} = T(t)\frac{T(h)z - z}{h}.$$

For $z \in \mathcal{D}(A)$ this tends to $T(t)Az$ as $h \to 0+$. Thus $T(t)z \in \mathcal{D}(A)$ and $AT(t)z = T(t)Az$, as asserted. □

The following example will be of interest to us later.

Example 2.2.5 *Let $\mathcal{H} = L^2(0, \infty)$, and let $(T(t))$ denote the* right shift semi-group, *defined by*

$$(T(t)f)(s) = \begin{cases} 0 & \text{for } s < t, \\ f(s - t) & \text{for } s \geq t. \end{cases}$$

Then, when it exists,

$$(Af)(s) = \lim_{h \to 0+} \frac{f(s - h) - f(s)}{h} = -f'(s),$$

so that

$$\mathcal{D}(A) = \{f \in L^2(0, \infty) : f \in AC, f' \in L^2(0, \infty)\},$$

where AC denotes the absolutely continuous functions (those that can be expressed as indefinite integrals of L^1 functions). By means of the Laplace transform we may pass to the unitarily equivalent semigroup $(\tilde{T}(t))$ on the Hardy space $\mathcal{K} = H^2(\mathbb{C}_+)$ of the right half-plane, defined by $(\tilde{T}(t)F)(s) = e^{-st}F(s)$ for $F \in H^2(\mathbb{C}_+)$. Its infinitesimal generator is $\tilde{A}$, which is defined on its domain by $(\tilde{A}F)(s) = -sF(s)$.

Even though the infinitesimal generator need not be a bounded operator, it is fairly well behaved, as the next result shows.

Theorem 2.2.6 *Let A be the infinitesimal generator of a C_0 semigroup $(T(t))$ defined on $\mathcal{X}$. Then A is a closed operator, and $\mathcal{D}(A)$ is dense in $\mathcal{X}$.*

Proof: To show that $\mathcal{D}(A)$ is dense, take $x \in X$; the approximants to x that we shall use are given by

$$x_t = \frac{1}{t} \int_0^t T(s)x \, ds, \quad t > 0,$$

where in general the integral of a continuous Banach space valued function $\phi : [a, b] \to \mathcal{X}$ can be simply defined as a limit of Riemann sums:

$$\int_a^b \phi(s) \, ds = \lim_{n \to \infty} \frac{1}{n} \sum_{k=1}^n \phi(a + k(b - a)/n).$$

It is clear that $x_t \to x$ as $t \to 0$; moreover, if $0 < h < t$, we have

$$
\begin{aligned}
\frac{T(h)x_t - x_t}{h} &= \frac{1}{th} \left(\int_h^{t+h} T(s)x \, ds - \int_0^t T(s)x \, ds \right) \\
&= \frac{1}{th} \left(\int_t^{t+h} - \int_0^h \right) T(s)x \, dx \to \frac{T(t)x - x}{t}
\end{aligned}
$$

as $h \to 0$. Thus $x_t \in \mathcal{D}(A)$, with

$$Ax_t = \frac{T(t)x - x}{t}, \tag{2.3}$$

and so $\mathcal{D}(A)$ is dense.

Now suppose that (x_n) is a sequence in $\mathcal{D}(A)$ with $x_n \to x$ and $Ax_n \to y$ as $n \to \infty$. We need to show that $x \in \mathcal{D}(A)$ and $y = Ax$. The first observation is that for $z \in \mathcal{D}(A)$ the derivative of the function $t \mapsto T(t)z$ exists and equals $T(t)Az$; indeed

$$\frac{T(t + h)z - T(t)z}{h} = T(t)\frac{T(h)z - z}{h} \to T(t)Az \qquad \text{as } h \to 0+,$$

and for $k > 0$ we have

$$\frac{T(t - k)z - T(t)z}{-k} = \frac{T(t - k)(z - T(k)z)}{-k} \to T(t)Az \qquad \text{as } k \to 0+,$$

since

$$\frac{T(t-k)(z-T(k)z)}{-k} - T(t)Az = (T(t-k)-T(t))\left(\frac{z-T(k)z}{-k}\right)$$
$$+ T(t)\left(\frac{z-T(k)z}{-k} - Az\right),$$

and both terms tend to zero as $k \to 0+$.

Thus, for $h > 0$ we have

$$\frac{T(h)z-z}{h} = \frac{1}{h}\int_0^h T(t)Az\,dt, \qquad \text{for } z \in \mathcal{D}(A). \tag{2.4}$$

Putting $z = x_n$ in (2.4) and letting $n \to \infty$, we obtain

$$\frac{T(h)x-x}{h} = \frac{1}{h}\int_0^h T(t)y\,dt.$$

Finally we let $h \to 0+$ to conclude that Ax exists and equals y. $\qquad\square$

We can also invert $A - \lambda I$ for λ in a suitable right half-plane, as the following result shows.

Proposition 2.2.7 *Let $(T(t))$ be a C_0 semigroup on a Banach space $\mathcal{X}$ with infinitesimal generator A, and suppose that $\|T(t)\| \le Me^{\alpha t}$ for all $t > 0$. Then $A - \lambda I$ is invertible for $\operatorname{Re}\lambda > \alpha$, in the sense that there is a bounded operator B_λ with $(A-\lambda I)B_\lambda x = x$ for all $x \in \mathcal{X}$ and $B_\lambda(A-\lambda I)x = x$ for all $x \in \mathcal{D}(A) = \mathcal{D}(A-\lambda I)$.*

Proof: Equation (2.3) with $T(t)$ replaced by $e^{-\lambda t}T(t)$ shows that

$$e^{-\lambda h}T(h)x - x = (A - \lambda I)\int_0^h e^{-\lambda t}T(t)x\,dt,$$

for $x \in \mathcal{X}$ and $h \ge 0$. (Note that $A - \lambda I$ is the infinitesimal generator of the semigroup $(e^{-\lambda t}T(t))$, as is easily verified.) Letting $h \to \infty$ and using the fact that A is closed, we see that

$$x = -(A - \lambda I)\int_0^\infty e^{-\lambda t}T(t)x\,dt.$$

Similarly, for $x \in \mathcal{D}(A)$ we obtain

$$x = -\int_0^\infty e^{-\lambda t}T(t)(A - \lambda I)x\,dt$$

by (2.4), and thus $(A - \lambda I)^{-1}$ exists and is given by

$$(A - \lambda I)^{-1}x = - \int_0^\infty e^{-\lambda t}T(t)x\,dt, \qquad (\operatorname{Re}\lambda > \alpha),$$

which we recognise as an operator-valued Laplace transform.   □

When we work in a Hilbert space, rather than a general Banach space, C_0 semigroups behave well under taking adjoints. The following result is perhaps not very surprising.

Theorem 2.2.8 *Let $(T(t))$ be a C_0 semigroup on a Hilbert space $\mathcal{H}$ with in-finitesimal generator A. Then $(T(t)^*)$ is also a C_0 semigroup, and its infinitesimal generator is A^*.*

Proof: We know by Corollary 2.1.9 that whenever A is a closed operator with dense domain, then A^* has the same property. Suppose that $\|T(t)\| \leq Me^{\alpha t}$ for all $t \geq 0$. Let us choose $x \in \mathcal{D}(A^*)$ and $y \in \mathcal{H}$, and take a real number λ with $\lambda > \alpha$. Then

$$\langle (e^{-\lambda h}T(h)^* - I)x, y \rangle = \langle x, (e^{-\lambda h}T(h) - I)y \rangle$$
$$= \left\langle (A^* - \lambda I)x, \int_0^h e^{-\lambda t}T(t)y\,dt \right\rangle,$$

as in Proposition 2.2.7. Therefore, taking the supremum over $\|y\| \leq 1$, we obtain

$$\|(e^{-\lambda h}T(h)^* - I)x\| \leq \|(A^* - \lambda I)x\|Mh \to 0 \qquad \text{as}\quad h \to 0.$$

It follows easily that $T(t)^*$ is also a C_0 semigroup. Moreover,

$$\left\langle \frac{(e^{-\lambda h}T(h)^* - I)x}{h}, y \right\rangle = \left\langle (A^* - \lambda I)x, \frac{1}{h}\int_0^h e^{-\lambda t}T(t)y\,dt \right\rangle \to \langle (A^* - \lambda I)x, y \rangle$$

as $h \to 0$, showing that $A^* - \lambda I$ is the infinitesimal generator of the semigroup $(e^{-\lambda t}T(t)^*)$, and hence A^* generates $(T(t)^*)$.   □

An operator T is said to be a *contraction* if $\|T\| \leq 1$, and a semigroup $(T(t))$ of operators is said to be a *contraction semigroup* if each $T(t)$ is a contraction. The following celebrated theorem, for which we merely sketch the proof, gives a necessary and sufficient condition for $(T(t))$ to be a contraction semigroup, in terms of the operator A.

Theorem 2.2.9 (Hille–Yosida) *Let A be an unbounded operator on a Banach space $\mathcal{X}$, with domain $\mathcal{D}(A)$. Then A is the infinitesimal generator of a C_0 contraction semigroup if and only if the following conditions hold:*

   1. A is a closed operator with dense domain.

2. $\{x \in \mathbb{R} : x > 0\} \subseteq \rho(A)$ *and* $\|(A - \lambda I)^{-1}\| \leq 1/\lambda$ *for all* $\lambda > 0$.

Proof: The idea is to define $A_n = -n^2(A - nI)^{-1} - nI = -nA(A - nI)^{-1}$ for $n \in \mathbb{N}$. These are bounded operators and generate semigroups $(T_n(t))$, defined by $T_n(t) = e^{A_n t}$ for $t \geq 0$.

It is then possible to verify that $A_n x \to x$ as $n \to \infty$ for each $x \in \mathcal{D}(A)$ and that $T(t)x := \lim_{n \to \infty} T_n(t)x$ exists and defines a C_0 semigroup whose infinitesimal generator is A.

Finally,

$$\|T_n(t)\| \leq e^{-nt} \exp(\|n^2(A - nI)^{-1}\|t) \leq e^{-nt}e^{nt} = 1$$

for each n and t, showing that $(T(t))$ is a contraction semigroup. $\square$

Consider now the following differential equation on an interval $[0, \infty)$ (it is also possible to consider a smaller interval $[0, \tau]$ instead):

$$\begin{aligned}
\frac{dx(t)}{dt} &= Ax(t), \qquad t > 0, \\
x(0) &= x_0.
\end{aligned} \tag{2.5}$$

If A is the infinitesimal generator of the C_0 semigroup $(T(t))$ on $\mathcal{X}$, then the function $t \mapsto x(t)$ is said to be a *mild solution* to (2.5) if $x_0 \in \mathcal{X}$ and $x(t) = T(t)x_0$.

Note that, as shown above, the derivative of $t \to T(t)x_0$ is indeed $T(t)Ax_0$, which is the same as $AT(t)x_0$, whenever $x_0 \in \mathcal{D}(A)$. Thus $t \mapsto x(t)$ is a *classical solution* to (2.5) if $x_0 \in \mathcal{D}(A)$ and $x(t) = T(t)x_0$. Again the special case $x(t) = e^{At}x_0$ for $A \in \mathcal{L}(\mathcal{X})$ clarifies what is going on here.

We digress now to introduce the concept of an *admissible observation operator*, since this, together with the associated *Weiss conjecture*, has stimulated a great deal of very recent research. To do this, we consider the following differential equation:

$$\begin{aligned}
\frac{dx(t)}{dt} &= Ax(t), \qquad x(0) = x_0, \\
y(t) &= Cx(t), \qquad t > 0.
\end{aligned} \tag{2.6}$$

In (2.6) we shall take A to be the infinitesimal generator of a C_0 semigroup $(T(t))$ on a Hilbert space $\mathcal{H}$ and $x_0 \in \mathcal{D}(A)$. The operator $C : \mathcal{D}(A) \to \mathcal{K}$ maps into another Hilbert space $\mathcal{K}$ (even the case $\mathcal{K} = \mathbb{C}$ is of interest here) and is called an *observation operator*, producing the *output* y. We may think of it as providing an observation of $x(t)$, the *state* of the evolving system. From the previous example, we know that the solution is $y(t) = CT(t)x_0$, provided that this expression makes sense. It is not necessary for C to be bounded with respect to the norm on $\mathcal{H}$ (if it was, it would even have a continuous extension to the whole of $\mathcal{H}$): it is sufficient that C be bounded with respect to the graph norm, and we can express this most clearly by requiring that there exist constants K_1, $K_2 > 0$ such that

$$\|Cz\| \leq K_1\|z\| + K_2\|Az\| \qquad \text{for all } z \in \mathcal{D}(A).$$

The operator C is said to be *(infinite-time) admissible*, if, informally, the output y is always square-integrable; more precisely, if there exists a constant $M > 0$ such that

$$\int_0^\infty \|CT(t)x_0\|^2 \, dt \leq M\|x_0\|^2 \qquad \text{for all } x_0 \in \mathcal{H}. \tag{2.7}$$

Now, checking whether a given observation operator C is admissible is not by any means straightforward: if, for example, A is given as a differential operator, then we may not be able to write down $T(t)$ very explicitly. However, if (2.7) is satisfied, then, taking the inner product of the function $t \mapsto CT(t)x_0$ with the function $t \mapsto e^{-\bar{s}t}$ for $s \in \mathbb{C}_+$ (i.e., $s \in \mathbb{C}$ with $\operatorname{Re} s > 0$) and using the Cauchy–Schwarz inequality $|\langle f, g \rangle| \leq \|f\| \, \|g\|$, we obtain the *resolvent condition* that there exists a constant $M' > 0$ such that

$$\|C(sI - A)^{-1}\| \leq \frac{M'}{\sqrt{\operatorname{Re} s}} \qquad \text{for all } s \in \mathbb{C}_+. \tag{2.8}$$

The Weiss conjecture was that (2.7) and (2.8) were equivalent. It turns out that for many semigroups, for example semigroups in which $T(t)$ is a normal operator for each t, and for all contraction semigroups, the conjecture holds for finite-dimensional $\mathcal{K}$. However, the conjecture does not hold for the class of all bounded semigroups and finite-dimensional $\mathcal{K}$; it also fails in general for infinite-dimensional $\mathcal{K}$. One of the features of this conjecture is that the cases in which it is known to be valid provide a simultaneous generalization of several celebrated results in the theory of Hardy spaces, Carleson measures, and Hankel operators. We do not go into further details here.

Example 2.2.10 As a further example of a C_0 semigroup, we consider the *heat semigroup*, which has been studied from a variety of points of view: analysis, probability, and of course applied mathematics and physics. We begin with the heat equation for a temperature distribution $x(r, t)$ on $[0, \infty)$, namely,

$$\frac{\partial x}{\partial t} = \frac{\partial^2 x}{\partial r^2}, \qquad x(r, 0) = x_0(r) \quad \text{given.} \tag{2.9}$$

Here r denotes position and t time, and we assume for the purposes of this example that the function $r \mapsto x(r, t)$ lies in $L^2(0, \infty)$ for each $t \geq 0$. Comparison with equation (2.5) suggests consideration of a semigroup $(T(t))$ with its infinitesimal generator A taken to be the differential operator $\partial^2/\partial r^2$: there are natural analogues in $\mathbb{R}^N$, where A becomes the Laplacian Δ given by

$$\Delta = \sum_{k=1}^{N} \frac{\partial^2}{\partial r_j^2}.$$

Again it is easier to analyse this situation in the Laplace domain, where we work on $H^2(\mathbb{C}_+)$ and the semigroup $(\tilde{T}(t))$ defined by $(\tilde{T}(t)F)(s) = \exp(ts^2)F(s)$, with infinitesimal generator $\tilde{A}$ satisfying $(\tilde{A}F)(s) = s^2 F(s)$. A solution to (2.9) is given by convolution with the *heat kernel*, that is,

$$(T(t)x_0)(r) = x(r,t) = \int_0^\infty \frac{1}{\sqrt{4\pi t}} e^{-(r-z)^2/4t} x_0(z)\, dz, \tag{2.10}$$

which corresponds to $\tilde{T}(t)$ by the Laplace transform.

The following final example is taken from [20].

Example 2.2.11 The displacement of a simply supported undamped beam may be modelled by the equation

$$\frac{\partial^2 y}{\partial t^2} = -\frac{\partial^4 y}{\partial x^4}, \qquad 0 \le x \le 1, \quad t \ge 0, \tag{2.11}$$

with initial conditions on the position and velocity,

$$y(x,0) = y_1(x) \qquad \text{and} \quad y_t(x,0) = y_2(x),$$

given, and boundary conditions

$$y(0,t) = y(1,t) = y_{xx}(0,t) = y_{xx}(1,t) = 0,$$

indicating that the beam is fixed at the endpoints.

We introduce the operator $B = -\dfrac{d^2}{dx^2}$ with domain

$$\mathcal{D}(B) = \{z \in L^2(0,1) : z, \frac{dz}{dx} \text{ abs. cont.}, \frac{d^2 z}{dx^2} \in L^2(0,1), z(0) = z(1) = 0\}.$$

We can then re-express (2.11) in the form

$$\frac{dz}{dt} = Az, \qquad \text{where}$$

$$z = \begin{pmatrix} y \\ \frac{dy}{dt} \end{pmatrix}, \qquad \text{and}$$

$$A = \begin{pmatrix} 0 & I \\ -B^2 & 0 \end{pmatrix},$$

where z lies in $\mathcal{D}(A)$, a subspace of the Hilbert space $Z = \mathcal{D}(B) \oplus L^2(0,1)$, equipped with the norm $\|(z_1, z_2)\|^2 = \|Bz\|^2 + \|z_2\|^2$.

This concludes our brief survey of semigroup theory, and we now introduce a technique for studying closed subspaces and closed operators on a Hilbert space.

2.3   The gap metric

We start by defining the notion of distance between two closed subspaces of a Hilbert space. Then, by using graphs, we shall be able to define the distance between two closed operators.

Definition 2.3.1 *Let V and W be closed subspaces of a Hilbert space $\mathcal{H}$. Then the* gap *between V and W, denoted by $\delta(V, W)$, is given by*

$$\delta(V, W) = \|P_V - P_W\|,$$

where P_V and P_W denote the orthogonal projections from $\mathcal{H}$ onto V and W, respectively.

With this definition it is easy to see that δ is a metric (see the exercises). However, some writers define the gap by a different formula, and we shall now show that the two are equivalent.

Theorem 2.3.2 *For two closed subspaces V and W of a Hilbert space $\mathcal{H}$ one has*

$$\delta(V, W) = \max\{\|(I - P_W)P_V\|, \|(I - P_V)P_W\|\}.$$

In particular, $0 \le \delta(V, W) \le 1$ for all V and W.

Proof: We begin with the following matrix identity, which is easily verified, bearing in mind that P_V and P_W are projections:

$$\begin{pmatrix} P_V \\ I - P_V \end{pmatrix} (P_V - P_W)(I - P_W \quad P_W) = \begin{pmatrix} P_V(I - P_W) & 0 \\ 0 & (I - P_V)P_W \end{pmatrix}. \tag{2.12}$$

The first factor on the left-hand side of (2.12) represents an isometric isomorphism between $V \times V^\perp$ and $\mathcal{H}$ and the third an isometric isomorphism between $\mathcal{H}$ and $W^\perp \times W$. Hence

$$
\begin{aligned}
\|P_V - P_W\| &= \left\| \begin{pmatrix} P_V(I - P_W) & 0 \\ 0 & (I - P_V)P_W \end{pmatrix} \right\|_{\mathcal{L}(W^\perp \times W, V \times V^\perp)} \\
&= \max\{\|P_V(I - P_W)\|, \|(I - P_V)P_W\|\},
\end{aligned}
$$

which is equivalent to the required result, since the adjoint of $P_V(I - P_W)$ is $(I - P_W)P_V$.   □

We write $\vec{\delta}(\mathcal{V}, \mathcal{W}) = \|(I - P_{\mathcal{W}})P_{\mathcal{V}}\|$ and call it the *directed gap* between $\mathcal{V}$ and $\mathcal{W}$. Since $\|(I - P_{\mathcal{W}})x\|$ is just the distance from x to $\mathcal{W}$, we see that

$$\vec{\delta}(\mathcal{V}, \mathcal{W}) = \sup_{x \in \mathcal{H}, \|x\|=1} \text{dist}(P_{\mathcal{V}}x, \mathcal{W}) = \sup_{v \in \mathcal{V}, \|v\|=1} \text{dist}(v, \mathcal{W}), \qquad (2.13)$$

at least when $\mathcal{V} \neq \{0\}$. Similar formulae hold with $\mathcal{V}$ and $\mathcal{W}$ interchanged.

There are problems when we attempt to define the gap for subspaces $\mathcal{V}$, $\mathcal{W}$ of a general Banach space $\mathcal{X}$, because there is no guarantee that bounded projections $P_{\mathcal{V}}$ and $P_{\mathcal{W}}$ exist (and certainly we cannot make sense of the idea of an orthogonal projection, such as we employ in a Hilbert space). We could still define $\delta(\mathcal{V}, \mathcal{W}) = \max\{\vec{\delta}(\mathcal{V}, \mathcal{W}), \vec{\delta}(\mathcal{W}, \mathcal{V})\}$, with $\vec{\delta}(\mathcal{V}, \mathcal{W}) = \sup_{v \in V, \|v\|=1} \text{dist}(v, \mathcal{W})$, but this is not always a metric. Instead we may use

$$\delta_1(\mathcal{V}, \mathcal{W}) = \max\{\vec{\delta}_1(\mathcal{V}, \mathcal{W}), \vec{\delta}_1(\mathcal{W}, \mathcal{V})\}, \qquad \text{where}$$
$$\vec{\delta}_1(\mathcal{V}, \mathcal{W}) = \sup_{v \in v, \|v\|=1} \text{dist}(v, S_{\mathcal{W}}),$$

writing $S_{\mathcal{W}} = \{w \in \mathcal{W} : \|w\| = 1\}$ for the unit sphere of $\mathcal{W}$. This is indeed a metric on the non-zero closed subspaces of $\mathcal{X}$. Further details can be found in [67].

It is perhaps a little hard to visualize what it means for the gap to be small, except in two or three dimensions (see the exercises for an example involving lines in the plane). However, we note that $\delta(\mathcal{V}, \mathcal{W}) = 1$ as soon as either space contains a vector orthogonal to the whole of the other space, for example, whenever one is a proper subspace of the other. We shall see some more interesting examples appear when we come to look at subspaces determined by operators.

It should now be clear from our previous discussions how to define the gap between two Hilbert space operators $A : \mathcal{D}(A) \to \mathcal{K}$ and $B : \mathcal{D}(B) \to \mathcal{K}$ when their domains $\mathcal{D}(A)$ and $\mathcal{D}(B)$ are subspaces of a common space $\mathcal{H}$ and their graphs $\mathcal{G}(A)$ and $\mathcal{G}(B)$ are closed subspaces of $\mathcal{H} \times \mathcal{K}$.

Definition 2.3.3 *Let A and B be as above. Then the gap between A and B, denoted $\delta(A, B)$, is defined by*

$$\delta(A, B) = \delta(\mathcal{G}(A), \mathcal{G}(B)).$$

Although δ is being used in two different senses, as a distance between subspaces and as a distance between operators, no confusion is likely. The analogous result for subspaces shows immediately that δ provides a metric on the set of closed operators between $\mathcal{H}$ and $\mathcal{K}$ and defines the *gap topology*.

For simplicity, we shall use P_T to denote the orthogonal projection onto the graph $\mathcal{G}(T)$ of T. When T is bounded, this can be written down explicitly, as follows.

Theorem 2.3.4 *Let $T : \mathcal{H} \to \mathcal{K}$ be a bounded operator; then the orthogonal projection $P_T : \mathcal{H} \times \mathcal{K} \to \mathcal{G}(T)$ is given by*

$$P_T \begin{pmatrix} h \\ k \end{pmatrix} = \begin{pmatrix} I \\ T \end{pmatrix} (I + T^*T)^{-1} \begin{pmatrix} I & T^* \end{pmatrix} \begin{pmatrix} h \\ k \end{pmatrix} \qquad (h \in \mathcal{H}, \ k \in \mathcal{K}).$$

Proof: By Theorem 2.1.8, we may write

$$(h, k) = (x, Tx) + (-T^*y, y),$$

uniquely, where $(x, Tx) \in \mathcal{G}(T)$ and $(-T^*y, y) \in \mathcal{G}(T)^\perp$. Thus we have

$$\begin{aligned} h &= x - T^*y, \\ k &= Tx + y. \end{aligned}$$

Hence $h + T^*k = (I + T^*T)x$, and the result is now clear. Note that $I + T^*T$ is a self-adjoint operator and if $\|x\| = 1$, then

$$\|(I + T^*T)x\| \geq |\langle (I + T^*T)x, x \rangle| = \|x\|^2 + \|Tx\|^2 \geq 1,$$

so that the inverse does indeed exist.   □

A slight generalization of this result can be found in Exercise 16.

Corollary 2.3.5 *The gap topology restricted to $\mathcal{L}(\mathcal{H}, \mathcal{K})$ gives the norm topology.*

Proof: Since both topologies are given by metrics, it is sufficient to check that a sequence converging in one metric also converges in the other (and vice versa). If (T_n) is a sequence in $L(\mathcal{H}, \mathcal{K})$ such that $\|T_n - T\| \to 0$ for some $T \in L(\mathcal{H}, \mathcal{K})$, then it is clear from Theorem 2.3.4 that $\|P_{T_n} - P_T\| \to 0$, so that the operators also converge in the graph metric.

Conversely, if $\|P_{T_n} - P_T\| \to 0$, then, taking $k = 0$ in Theorem 2.3.4, we have $(I + T_n^*T_n)^{-1} \to (I + T^*T)^{-1}$ in norm, and hence $I + T_n^*T_n \to I + T^*T$ in norm. We also have $T_n(I + T_n^*T_n)^{-1} \to T(I + T^*T)^{-1}$ in norm and so, multiplying convergent sequences, we see that $T_n \to T$ in norm.   □

Notes

Some standard introductory texts in this area of operator theory are [9, 116, 117, 146].

Closed operators are discussed further in [25, 67, 146], for example.

More advanced accounts of the theory of semigroups can be found in the monographs [22, 23, 25, 30, 56, 67, 106]. For the ideas of admissibility and the Weiss conjecture we refer the reader to the articles [63, 60, 65, 141, 143].

Our proof of the formula for the gap metric is adapted from [40], which treats a special case. See also [31]. There are other, longer, proofs in the literature, for example, that in [67].

Exercises

1. Show that if $T : \mathcal{X} \to \mathcal{Y}$ is a linear mapping between vector spaces, then $\mathcal{G}(T)$ is a linear subspace of the vector space $\mathcal{X} \times \mathcal{Y}$.

2. Let $\mathcal{H}$ and $\mathcal{K}$ be complex inner-product spaces, and for vectors $v_1 = (h_1, k_1)$ and $v_2 = (h_2, k_2)$ in $\mathcal{H} \times \mathcal{K}$ define $\langle v_1, v_2 \rangle = \langle h_1, h_2 \rangle + \langle k_1, k_2 \rangle$. Show that this makes $\mathcal{H} \times \mathcal{K}$ into an inner-product space and that the induced norm is given by (2.1) with $p = 2$.

3. Let $\mathcal{X}$ be an infinite-dimensional normed space, and suppose that $S = \{x_\lambda : \lambda \in \Lambda\}$ is a *Hamel basis* of $\mathcal{X}$, that is, a maximal set of vectors such that all finite subsets are linearly independent (these can be shown to exist by using Zorn's lemma). Show that every vector is a linear combination of vectors in S in a unique way, and that for any vector space $\mathcal{Y}$ we can define a unique linear mapping $T : \mathcal{X} \to \mathcal{Y}$ by specifying Tx_λ arbitrarily for $x_\lambda \in S$. Suppose now that $\mathcal{Y} \neq \{0\}$ is any normed space. Show that there are unbounded but everywhere-defined linear mappings from $\mathcal{X}$ to $\mathcal{Y}$.

4. Let T be the operator given in Example 2.1.3, which is densely defined and unbounded. Calculate its adjoint T^* and specify $\mathcal{D}(T^*)$.

5. Verify that the infinitesimal generator of a C_0 semigroup is a linear operator and that its domain of definition is a linear subspace.

6. Let A be a bounded operator on a Banach space $\mathcal{X}$. Show that we may define $T(t) = \exp(At)$ by means of a norm-convergent power series, that $(T(t))$ is a semigroup such that $t \mapsto T(t)$ is norm-continuous, and that A is

the infinitesimal generator of the semigroup $(T(t))$.

7. Let $(\lambda_n)_{n=1}^{\infty}$ be a sequence of complex numbers and define a semigroup $(T(t))$ on a Hilbert space $\mathcal{H}$ with orthonormal basis $(e_n)_{n=1}^{\infty}$ by

$$T(t) \sum_{n=1}^{\infty} a_n e_n = \sum_{n=1}^{\infty} e^{\lambda_n t} a_n e_n,$$

where $\sum_{n=1}^{\infty} a_n e_n$ is a vector in $\mathcal{H}$. Show that, provided the sequence $(\operatorname{Re} \lambda_n)$ is bounded above, the operators $T(t)$ are bounded and form a C_0 semigroup. What is its infinitesimal generator?

8. Suppose that $(T(t))$ is a C_0 semigroup with infinitesimal generator A. Take $\lambda \in \mathbb{C}$. Show that $(e^{-\lambda t} T(t))$ is also a C_0 semigroup and that its infinitesimal generator is $A - \lambda I$.

9. Let $\mathcal{H}$ be a Hilbert space with orthonormal basis (e_n), and let $(T(t))$ be the C_0 semigroup on $\mathcal{H}$ with infinitesimal generator A satisfying $Ae_n = a_n e_n$ for each n, where (a_n) is an arbitrary real sequence. Show that $(T(t))$ is a normal semigroup, so that the resolvent condition (2.8) is equivalent to admissibility. Suppose that the observation functional $C : \mathcal{D}(A) \to \mathbb{C}$ is given by $Ce_n = c_n$, where (c_n) is also a real sequence. Write down (2.8) explicitly in terms of (a_n) and (c_n).

10. Verify directly that (2.10) satisfies the heat equation.

11. Check that the gap metric δ is indeed a metric on the collection of closed subspaces of $\mathcal{H}$. (Note that $P_V = P_W$ if and only if $V = W$. Why?)

12. Prove that $\delta(V, W) = \delta(V^{\perp}, W^{\perp})$ for all subspaces $V, W \subseteq \mathcal{H}$.

13. Give a non-matricial proof of Theorem 2.3.2, starting from the identity $P_V - P_W = (I - P_W)P_V - P_W(I - P_V)$.

14. Let $p \in \mathbb{C} \setminus \{0\}$ be fixed. Calculate the gap between the one-dimensional subspaces $V_0 = \{(x, 0) : x \in \mathbb{C}\}$ and $V_m = \{(x, px) : x \in \mathbb{C}\}$ of the two-dimensional Hilbert space $\mathbb{C}^2$, sometimes written ℓ_2^2. This can be done

either by writing down the projections explicitly and using the definition of the gap or, alternatively, by calculating the two directed gaps $\vec{\delta}$.

15. Show that the Banach space gap δ_1 is a metric. Repeat the calculation of Exercise 14 using δ_1 in place of δ.

16. Suppose that $T : \mathcal{H} \to \mathcal{H}$ is a closed densely-defined operator with graph $\mathcal{G}(T) = \{(Ah, Bh) : h \in \mathcal{H}\}$ for a pair A, B of commuting bounded operators on $\mathcal{H}$. Find an explicit formula for the orthogonal projection $P_T : \mathcal{H} \times \mathcal{H} \to \mathcal{G}(T)$.

17. Given T_n, T as in the proof of Corollary 2.3.5, derive explicit upper bounds for $\|P_{T_n} - P_T\|$ in terms of $\|T_n - T\|$, and vice versa.

Chapter 3

Shift-invariance and causality

Our philosophy is to regard a linear system as a linear operator with the two additional properties of shift-invariance (also known as time-invariance) and causality – in fact, time-varying systems are also of importance, but we shall concentrate on the time-invariant case. Such an operator will have a graph that is a shift-invariant subspace, and so, apart from their general importance in operator theory, it is important for us to understand the nature of these subspaces.

Let us first present an example. The operator of multiplication by the function $s \mapsto \dfrac{e^{-sh}}{s+a}$ (with $a \in \mathbb{R}$ and $h > 0$) on the Hardy space $H^2(\mathbb{C}_+)$ can be naturally identified with the delay-differential equation

$$\frac{dy}{dt} + ay = u(t - h),$$

a so-called *dead-time system* (see [88]), which occurs in a variety of contexts, including, for example, the catalytic convertor in a car engine [12]. We shall see a more detailed treatment of delay equations in Chapter 6, but meanwhile we shall see in this chapter that in some sense "all" linear time-invariant systems correspond to operators of multiplication.

3.1 Invariant subspaces

Suppose that T is a bounded linear operator on a complex Hilbert space $\mathcal{H}$; then an *invariant subspace* for T is a linear subspace $\mathcal{K} \subseteq \mathcal{H}$ such that $Tk \in \mathcal{K}$ for all $k \in \mathcal{K}$. It is a very old question, unsolved at the time of writing, whether T necessarily possesses any closed invariant subspaces, apart from the obvious examples $\{0\}$ and $\mathcal{H}$ itself (see [16] for a recent survey of this question). The corresponding question for Banach spaces is known to have a negative answer in general, even on the space ℓ^1 [3, 29, 113, 114].

A fundamental operator that will be of interest to us is the *right shift*, R : $\ell^2(\mathbb{Z}_+) \to \ell^2(\mathbb{Z}_+)$, defined by

$$R(a_0, a_1, \ldots) = (0, a_0, a_1, \ldots), \qquad (a_n)_0^\infty \in \ell^2(\mathbb{Z}_+).$$

(We shall reserve the letter S for an equivalent Hardy space operator, to be defined shortly.) This operator has a natural extension to $\ell^2(\mathbb{Z})$, which we shall also denote by R, defined by

$$R(a_n)_{-\infty}^\infty = (b_n)_{-\infty}^\infty, \qquad \text{where } b_n = a_{n-1}, \quad n \in \mathbb{Z}.$$

Note that R is a unitary operator on $\ell^2(\mathbb{Z})$, but that R is not even a normal operator on $\ell^2(\mathbb{Z}_+)$ (see Exercise 10 of Chapter 2). For these two operators a complete and very elegant classification of their closed invariant subspaces is known, which we shall now derive. We then look at the analogous results on $L^2(0, \infty)$ and $L^2(\mathbb{R})$, as well as vector-valued cases. Part of our motivation in all this is to understand linear shift-invariant systems, which we shall regard as operators with closed shift-invariant graphs (usually there will also be a causality condition, to be discussed later).

By means of the unitary correspondence between $\ell^2(\mathbb{Z})$ and $L^2(\mathbb{T})$, under which $\ell^2(\mathbb{Z}_+)$ corresponds to the Hardy space $H^2 = H^2(\mathbb{D})$ (see Chapter 1), we may consider the problem as one of determining the closed invariant subspaces of the operator S of multiplication by z on $L^2(\mathbb{T})$, given more precisely by the formula

$$(Sf)(e^{i\omega}) = e^{i\omega} f(e^{i\omega}), \qquad f \in L^2(\mathbb{T}),$$

and its restriction to H^2.

It is convenient here to refer to operators on $\ell^2(\mathbb{Z})$ and $\ell^2(\mathbb{Z}_+)$ as acting in the *time domain* and in *discrete time*. The equivalent operators on $L^2(\mathbb{T})$ and $H^2(\mathbb{D})$ then act in the *frequency domain*.

There are two types of closed invariant subspaces that we may encounter. If $S\mathcal{K} \subseteq \mathcal{K}$, we may have either (i) $S\mathcal{K} \neq \mathcal{K}$ or (ii) $S\mathcal{K} = \mathcal{K}$. In the first case, $\mathcal{K}$ is called *1-invariant*, or *simply invariant*. In the second case, $\mathcal{K}$ is *2-invariant* or *doubly invariant*, and it is clear that $S^{-1}\mathcal{K} = \mathcal{K}$ as well.

Let us begin by classifying the 2-invariant subspaces. For a subset E of $\mathbb{T}$, we write χ_E for the characteristic function of E, so that

$$\chi_E(z) = \begin{cases} 1 & \text{if } z \in E, \\ 0 & \text{otherwise.} \end{cases}$$

Then $\chi_E L^2(\mathbb{T}) = \{f \in L^2(\mathbb{T}) : f = 0 \text{ a.e. on } \mathbb{T} \setminus E\}$. These functions occur in the following classification theorem.

Theorem 3.1.1 (Wiener) *A closed subspace* $\mathcal{K}$ *of* $L^2(\mathbb{T})$ *satisfies* $S\mathcal{K} = \mathcal{K}$ *if and only if* $\mathcal{K} = \chi_E L^2(\mathbb{T})$ *for some measurable subset* $E \subseteq \mathbb{T}$.

Proof: Write ϕ for the orthogonal projection of the constant function 1 on $\mathcal{K}$. Since $1 - \phi \in \mathcal{K}^\perp$ and $S^n\phi \in \mathcal{K}$ for all $n \in \mathbb{Z}$, we have $\langle S^n\phi, 1 - \phi \rangle = 0$ for all n; but this says that

$$\frac{1}{2\pi} \int_0^{2\pi} (\phi(e^{i\omega}) - |\phi(e^{i\omega})|^2)e^{in\omega}\, d\omega = 0 \qquad \text{for all } n \in \mathbb{Z},$$

and hence $\phi = |\phi|^2$ almost everywhere, and $\phi(e^{i\omega}) = 0$ or 1 a.e. Let E be the set on which $\phi = 1$, so $\phi = \chi_E$. We need to show that $\mathcal{K} = \chi_E L^2(\mathbb{T})$. Clearly we already have $\chi_E L^2(\mathbb{T}) \subseteq \mathcal{K}$.

Consider any function $f \in \mathcal{K}$ with $f \perp \chi_E L^2(\mathbb{T})$; this implies that

$$\frac{1}{2\pi} \int_0^{2\pi} f(e^{i\omega})\chi_E(e^{i\omega})e^{-in\omega}\, d\omega = 0 \qquad \text{for all } n \in \mathbb{Z},$$

and we conclude that $f\chi_E = 0$. Now $S^n f \in \mathcal{K}$ and $1 - \chi_E \in \mathcal{K}^\perp$, and so

$$\frac{1}{2\pi} \int_0^{2\pi} e^{in\omega} f(e^{i\omega})(1 - \chi_E(e^{i\omega}))\, d\omega = 0 \qquad \text{for all } n \in \mathbb{Z}.$$

Thus $f(1 - \chi_E) = 0$ a.e., and finally $f = 0$. Thus $\mathcal{K} = \chi_E L^2(\mathbb{T})$, as asserted. $\square$

Note that no non-trivial subspace $\chi_E L^2(\mathbb{T})$ can be contained in H^2, by Theorem 1.2.3. This agrees with the easy observation that H^2 has no 2-invariant subspaces for S.

The classification of 1-invariant subspaces of $L^2(\mathbb{T})$ is slightly more complicated and is given by the *Beurling–Helson theorem*, which now follows. In this case, the result for H^2 is rather more interesting.

Theorem 3.1.2 *A closed subspace* $\mathcal{K}$ *of* $L^2(\mathbb{T})$ *satisfies* $S\mathcal{K} \subset \mathcal{K}$ *with* $S\mathcal{K} \neq \mathcal{K}$, *if and only if* $\mathcal{K} = \phi H^2$ *for some measurable function* ϕ *satisfying* $|\phi(e^{i\omega})| = 1$ *almost everywhere, unique to within a constant factor of modulus 1.*

Proof: We leave the reader to verify that if $|\phi| = 1$ a.e. on $\mathbb{T}$, then $\mathcal{K} = \phi H^2$ is a closed subspace of $L^2(\mathbb{T})$ and that $S\mathcal{K} \subset \mathcal{K}$, $S\mathcal{K} \neq \mathcal{K}$.

Let us now establish the uniqueness. If ϕ_1 and ϕ_2 are two functions, unimodular on the circle, such that $\phi_1 H^2 = \phi_2 H^2$, then $\phi_1 = f\phi_2$ and $\phi_2 = g\phi_1$ for two functions $f, g \in H^2$, which must themselves be unimodular, and hence inner. By the maximum principle, $|f| \leq 1$ in $\mathbb{D}$, and the same holds for $|g|$, and we conclude that f and g are constant.

Now, given such a 1-invariant subspace $\mathcal{K}$, we have $\mathcal{K} \neq S\mathcal{K}$, and so we can find a function $\phi \in \mathcal{K}$, which we may take to have unit norm, such that $\phi \perp S\mathcal{K}$. In particular $\phi \perp S^n\phi$ for $n = 1, 2, \ldots$. This gives

$$\frac{1}{2\pi} \int_0^{2\pi} |\phi(e^{i\omega})|^2 e^{-in\omega}\, d\omega = 0 \qquad \text{for all } n \geq 1; \tag{3.1}$$

but, by conjugation, this will also hold for all $n \leq -1$, and hence $|\phi|^2$ is a constant, and the constant is 1, because $\|\phi\| = 1$.

We may now exploit Theorem 3.1.1 and note that the sequence $(S^n\phi)_{n\in\mathbb{Z}}$ is orthonormal and that its closed linear span is the whole of $L^2(\mathbb{T})$, since there is no non-trivial subset of the circle on which all the functions vanish. Now the closed linear span of $(S^n\phi)_{n=0}^\infty$, which is ϕH^2, is contained in $\mathcal{K}$. Moreover, for $k \in \mathcal{K}$ and $n \geq 1$, we have

$$\langle S^{-n}\phi, k \rangle = \langle \phi, S^n k \rangle = 0,$$

since $\phi \perp S\mathcal{K}$ by construction. Thus $S^{-n}\phi \in \mathcal{K}^\perp$. We have thus established that $\phi H^2 \subseteq \mathcal{K}$ and $(\phi H^2)^\perp \subseteq \mathcal{K}^\perp$, and so we have equality in both cases. $\qquad\square$

The following corollary, the classical Beurling theorem, is immediate.

Corollary 3.1.3 (Beurling) *Let $\mathcal{K}$ be a non-zero subspace of H^2 that is invariant under S. Then $\mathcal{K} = \phi H^2$ for some inner function ϕ, which is unique to within a constant of modulus 1.*

This in turn allows us, given the inner–outer factorization described in Theorem 1.3.3, to find a much more user-friendly characterization of outer functions.

Corollary 3.1.4 *A function $u \in H^\infty$ is outer if and only if uH^2 is dense in H^2.*

Proof: For any $u \in H^\infty$, the closure of uH^2 is a closed subspace of H^2, and hence of the form ϕH^2 for some inner function ϕ. Evidently $u = \phi v$ for some $v \in H^\infty$. Now, if u is outer, it has no nontrivial inner divisors, and so ϕ is a constant and uH^2 is dense in H^2. Conversely, if u is not outer, then $uH^2 \subseteq \phi H^2$, where ϕ is a non-constant inner function dividing u, and so uH^2 is not dense in H^2. $\qquad\square$

Returning to our general discussion of invariant subspaces for bounded operators, it is easy to see that if $T \in \mathcal{L}(\mathcal{H})$ and $x \in \mathcal{H}$, then the closed linear span of $(T^n x)_{n=0}^\infty$ is a closed T-invariant subspace of $\mathcal{H}$. If this is the whole of $\mathcal{H}$, then x is said to be a *cyclic vector*, and the invariant subspace problem is equivalent to the question of whether every non-zero vector is cyclic. We have just seen that the cyclic vectors for $S \in \mathcal{L}(H^2)$ are the outer functions.

Before moving onto the vector-valued case, let us prove what is commonly known as the *Beurling–Lax theorem* about invariant subspaces of $L^2(0, \infty)$ invariant under the set $\{R_\lambda : \lambda \geq 0\}$ of all right shifts:

$$(R_\lambda f)(t) = \begin{cases} f(t - \lambda) & \text{if } t \geq \lambda, \\ 0 & \text{otherwise.} \end{cases}$$

By means of the (up to a constant) unitary equivalence $L : L^2(0, \infty) \mapsto H^2(\mathbb{C}_+)$ induced by the Laplace transform L (see Chapter 1), which satisfies $L(R_\lambda f)(s) = e^{-\lambda s}(Lf)(s)$, we may reduce the problem to finding the shift-invariant subspaces of $H^2(\mathbb{C}_+)$, that is, those invariant under the operators S_λ of multiplication by the function $s \mapsto e^{-\lambda s}$ for all $\lambda \geq 0$.

Again, we think of operators on $L^2(\mathbb{R})$ and $L^2(0, \infty)$ as acting in the *time domain* and their equivalents on $L^2(i\mathbb{R})$ and $H^2(\mathbb{C}_+)$ as acting in the *frequency domain*. This is now the *continuous-time* situation.

Theorem 3.1.5 *A non-zero closed subspace $\mathcal{K} \subseteq H^2(\mathbb{C}_+)$ satisfies $S_\lambda \mathcal{K} \subseteq \mathcal{K}$ for all $\lambda \geq 0$ if and only if $\mathcal{K} = \phi H^2(\mathbb{C}_+)$ for some inner function $\phi \in H^\infty(\mathbb{C}_+)$.*

Proof: Clearly every such subspace $\phi H^2(\mathbb{C}_+)$ is indeed closed and shift-invariant.

Let $V : H^2(\mathbb{D}) \mapsto H^2(\mathbb{C}_+)$ denote the isometric isomorphism described in Theorem 1.2.5; V is induced by the conformal bijection M defined by $M(s) = (1 - s)/(1 + s)$. It is sufficient to show that the closed subspace $V^{-1}\mathcal{K}$ of H^2 is invariant under S. Since $(VSV^{-1}f)(s) = M(s)f(s)$ for $f \in H^2(\mathbb{C}_+)$, it is enough to show that $\mathcal{K}$ is invariant under multiplication by the function

$$s \mapsto \frac{1 - s}{1 + s} = -1 + \frac{2}{1 + s}.$$

But the operator $\tilde{S}$ of multiplication by $1/(1 + s)$ can be approximated by combinations of the S_λ. Specifically,

$$\frac{1}{1 + s} = \int_0^\infty e^{-(1+s)t}\, dt = \lim_{n \to \infty} \int_0^n e^{-(1+s)t}\, dt$$

uniformly in s. On approximating this last integral by Riemann sums, we see that $1/(1 + s)$ is a bounded pointwise limit of linear combinations of functions $e^{-\lambda s}$. This, combined with Lebesgue's bounded convergence theorem, shows that for all $f \in \mathcal{K}$ we have $\tilde{S}f = \lim_{m \to \infty} S_m f$, where each $\tilde{S}_m$ is a finite combination of the S_λ. Hence $\tilde{S}f \in \mathcal{K}$, and this completes the proof. $\square$

We now move on to a discussion of the vector-valued case and will prove just two more theorems here. One is a generalization of the Wiener theorem on 2-invariant subspaces of $L^2(\mathbb{T}, \mathbb{C}^m)$, and the other is a generalization of the Beurling theorem on 1-invariant subspaces of $H^2(\mathbb{D}, \mathbb{C}^m)$. To avoid a proliferation of very similar results, we shall not derive the analogous half-plane results explicitly, but they are similar.

Instead of a function taking values in the set $\{0, 1\}$ almost everywhere, we now need to deal with functions whose values are orthogonal projections. Accordingly, we say that $P : \mathbb{T} \to \mathcal{L}(\mathbb{C}^m)$ is a *measurable projection-valued function* if it satisfies the following:

- $P(e^{i\omega})$ is the orthogonal projection onto some closed subspace $\mathcal{J}(e^{i\omega})$ of $\mathbb{C}^m$ for almost all $e^{i\omega} \in \mathbb{T}$.

- The mappings $\omega \mapsto \langle P(e^{i\omega})x, y \rangle$ are measurable for every $x, y \in \mathbb{C}^m$.

Since $P(e^{i\omega})$ can be regarded as an $m \times m$ matrix-valued function, the second property just says that $P \in L^\infty(\mathbb{T}, \mathcal{L}(\mathbb{C}^m))$. We can equally regard P as a projection operator on $L^2(\mathbb{C}^m, \mathbb{T})$, by pointwise multiplication. Note that

$$\operatorname{Im} P = \{f \in L^2(\mathbb{T}, \mathbb{C}^m) : f(e^{i\omega}) \in \mathcal{J}(e^{i\omega}) \text{ a.e.}\}.$$

In the case $m = 1$, an orthogonal projection on $\mathbb{C}$ is just a mapping $z \mapsto pz$, where $p^2 = p$, that is, $p = 0$ or 1. It will thus be easy to recover Theorem 3.1.1 from its vectorial counterpart, which we now give.

Theorem 3.1.6 *A closed subspace $\mathcal{K}$ of $L^2(\mathbb{T}, \mathbb{C}^m)$ satisfies $S\mathcal{K} = \mathcal{K}$ if and only if $\mathcal{K} = PL^2(\mathbb{T}, \mathbb{C}^m)$ for some measurable projection-valued function $P : \mathbb{T} \to \mathcal{L}(\mathbb{C}^m)$, which is unique to within sets of measure 0.*

Proof: It is easily verified that any subspace $\mathcal{K} = PL^2(\mathbb{T}, \mathbb{C}^m)$ is shift-invariant; it is also closed, since if a sequence in $PL^2(\mathbb{T}, \mathbb{C}^m)$ converges to f in norm, then it has a subsequence converging almost everywhere, and we easily conclude that $f(e^{i\omega}) \in \mathcal{J}(e^{i\omega})$ a.e.

Given such a closed shift-invariant subspace $\mathcal{K}$, let $P_\mathcal{K} : L^2(\mathbb{T}, \mathbb{C}^m) \to \mathcal{K}$ denote the orthogonal projection onto it. Let $\{e_1, \ldots, e_m\}$ denote the canonical basis of $\mathbb{C}^m$, and for $r \in \mathbb{Z}$ and $1 \le k \le m$ let $\phi_{r,k}$ denote the vector-valued function defined on $\mathbb{T}$ by $z \mapsto z^r e_k$. For each $\omega \in \mathbb{T}$ let $\mathcal{J}(e^{i\omega})$ denote the linear span (which is necessarily closed) of the set of vectors

$$\{(P_\mathcal{K}\phi_{r,k})(e^{i\omega}) : r \in \mathbb{Z}, 1 \le k \le m\}$$

in $\mathbb{C}^m$. This is well defined to within a set of measure 0. Define $P(e^{i\omega})$ to be the orthogonal projection from $\mathbb{C}^m$ onto $\mathcal{J}(e^{i\omega})$. It is clear that

$$\mathcal{K} \subseteq \{f \in L^2(\mathbb{T}, \mathbb{C}^m) : f(e^{i\omega}) \in \mathcal{J}(e^{i\omega}) \text{ a.e.}\}.$$

If the two spaces were unequal, then there would exist $f \in L^2(\mathbb{T}, \mathbb{C}^m)$ such that $f(e^{i\omega}) \in \mathcal{J}(e^{i\omega})$ a.e. but $f \perp S^n P_\mathcal{K} \phi_{r,k} = 0$ for all $n \in \mathbb{Z}$, $r \in \mathbb{Z}$ and $1 \le k \le m$. Thus

$$\frac{1}{2\pi} \int_0^{2\pi} \langle f(e^{i\omega}), e^{in\omega} \phi_{r,k}(e^{i\omega}) \rangle \, d\omega = 0$$

for each n and so $\langle f(e^{i\omega}), \phi_{r,k}(e^{i\omega}) \rangle = 0$ a.e. Thus $f(e^{i\omega}) = 0$ a.e., because it is a vector in $\mathcal{J}(e^{i\omega})$ that is orthogonal to a spanning set of $\mathcal{J}(e^{i\omega})$.

Finally, P is indeed measurable since if $v \in \mathbb{C}^m$ and we write $\tilde{v}$ for the function constantly equal to v, then $P_\mathcal{K} \tilde{v} \in \mathcal{K} \subseteq L^2(\mathbb{T}, \mathbb{C}^m)$, and so $\omega \mapsto P(e^{i\omega})v$ is measurable. □

In order to state the Beurling–Lax theorem on $H^2(\mathbb{C}^m)$ we need the concept of an *operator-valued inner function*. This is a function $\phi \in H^\infty(\mathbb{T}, \mathcal{L}(\mathbb{C}^r, \mathbb{C}^m))$ satisfying the additional condition that $\phi(e^{i\omega})$ is an isometry a.e. on $\mathbb{T}$.

Theorem 3.1.7 *Let $\mathcal{K}$ be a non-zero subspace of $H^2(\mathbb{C}^m)$ that is invariant under S. Then there is an r with $0 \le r \le m$ and an inner function ϕ belonging to $H^\infty(\mathbb{T}, L(\mathbb{C}^r, \mathbb{C}^m))$ such that $\mathcal{K} = \phi H^2(\mathbb{C}^r)$.*

Proof: As in the proof of Theorem 3.1.2, we look at the difference between $\mathcal{K}$ and $S\mathcal{K}$, writing $\mathcal{W} = \mathcal{K} \ominus S\mathcal{K}$. Since $\mathcal{K} \subseteq H^2(\mathbb{C}^m)$, we see by elementary linear algebra that the dimension of W is at most m. Note that the subspaces $\mathcal{W}$, $S\mathcal{W}$, $S^2\mathcal{W}$, ... are pairwise orthogonal, since S is an isometry and so $\langle S^j w, S^k w' \rangle = \langle w, S^{k-j} w' \rangle = 0$ whenever $k > j$ and w, $w' \in \mathcal{W}$. Indeed, we claim that $\mathcal{K}$ is the orthogonal direct sum,

$$\mathcal{K} = \mathcal{W} \oplus S\mathcal{W} \oplus S^2\mathcal{W} \oplus \ldots; \tag{3.2}$$

for if $x \in \mathcal{K}$ and $x \perp \mathcal{W} \oplus S\mathcal{W} \oplus S^2\mathcal{W} \oplus \ldots$, then

- $x \perp \mathcal{W}$, so $x \in S\mathcal{K}$, and $x = Sx_1$ for some $x_1 \in \mathcal{K}$.
- $x = Sx_1 \perp S\mathcal{W}$, so $x_1 \perp \mathcal{W}$, so $x_1 \in S\mathcal{K}$, and $x_1 = Sx_2$ for some $x_2 \in \mathcal{K}$.
- $x = S^2 x_2 \perp S^2\mathcal{W}$, so $x_2 \perp \mathcal{W}$, ...

Continuing in this way, we see that x can be written as $x = S^n x_n$, with $x_n \in \mathcal{K}$, for each $n \in \mathbb{N}$. Since $x \in H^2(\mathbb{C}^m)$, we must have $x = 0$.

Any function $w \in \mathcal{W}$ has constant modulus a.e., since $w \perp S^n w$, for $n \ge 1$, so that

$$\frac{1}{2\pi} \int_0^{2\pi} \langle w(e^{i\omega}), e^{in\omega} w(e^{i\omega}) \rangle \, d\omega = 0,$$

and, by conjugation, we see that all the non-zero Fourier coefficients of $\omega \mapsto \langle w(e^{i\omega}), w(e^{i\omega}) \rangle$ are zero. (It is instructive at this point to compare Equation (3.1).)

Suppose that $\dim W = r \leq m$, let $\gamma : \mathbb{C}^r \to W$ be an isometric isomorphism, and for $z \in \mathbb{D}$ define $\phi(z) \in \mathcal{L}(\mathbb{C}^r, \mathbb{C}^m)$ by

$$\phi(z)v = \gamma(v)(z), \qquad \text{for all } v \in \mathbb{C}^r;$$

this is analytic as a function of z. Any function in $H^2(\mathbb{C}^r)$ can be written as $f(z) = \sum_{k=0}^{\infty} v_k z^k$, where $v_k \in \mathbb{C}^r$ and $\sum_{k=0}^{\infty} \|v_k\|^2 < \infty$. We see that ϕf is the same, pointwise, as $\sum_{k=0}^{\infty} S^k \gamma(v_k)$, and since

$$\|f\|^2 = \sum_{k=0}^{\infty} \|v_k\|^2 = \sum_{k=0}^{\infty} \|S^k \gamma(v_k)\|^2 = \|\phi f\|^2$$

by (3.2), we conclude finally that $\phi \in H^\infty(\mathbb{D}, L(\mathbb{C}^r, \mathbb{C}^m))$ is inner. $\qquad \square$

Remark 3.1.8 Probably the most general form of the Beurling–Lax theorem can be found in [115], which defines a shift operator on a general Hilbert space $\mathcal{H}$ as an operator S such that S is an isometry and $\|S^{*n}x\| \to 0$ for all $x \in \mathcal{H}$. (This includes the usual shift on $H^2(\mathbb{D})$ but not the shift on $L^2(\mathbb{T})$, which is unitary.) An operator A is said to be S-inner if $SA = AS$ and A is a partial isometry (which means that $\mathcal{H}$ splits as $\mathcal{H} = \mathcal{K} \oplus \mathcal{K}^\perp$, where $\mathcal{K} = \ker A$ and A is isometric on $\mathcal{K}^\perp$). Then it turns out that the closed S-invariant subspaces of $\mathcal{H}$ all have the form $A\mathcal{H}$, where A is S-inner. The reader will see better how this fits in with Theorems 3.1.2 and 3.1.7 after we have looked at shift-invariant operators in the next section.

3.2   Invariant operators

In this section we are interested in linear shift-invariant operators, by which we mean operators whose graphs are closed shift-invariant subspaces. Note that if T is a bounded operator on $\ell^2(\mathbb{Z}_+)$ such that $RT = TR$, then if $(x, Tx) \in \mathcal{G}(T)$, we have

$$R(x, Tx) = (Rx, TRx) \in \mathcal{G}(T),$$

and thus $\mathcal{G}(T)$ is a closed shift-invariant subspace of $\ell^2(\mathbb{Z}_+, \mathbb{C}^2)$. The converse also holds, for if $(Rx, RTx) \notin \mathcal{G}(T)$ for some $x \in \ell^2(\mathbb{Z}_+)$, then we do not have $RT = TR$.

Thus it makes sense to discuss closed shift-invariant operators on $\ell^2(\mathbb{Z}_+)$ and $L^2(0, \infty)$ (and, briefly, the analogous operators on $\mathbb{Z}$ and $\mathbb{R}$) by means of their graphs; by virtue of the results in Section 3.1, we shall find it simpler to proceed by transforming to the equivalent spaces, which are, respectively, $H^2(\mathbb{D})$, $H^2(\mathbb{C}_+)$, $L^2(\mathbb{T})$ and $L^2(i\mathbb{R})$. As before, we shall also look at the vector-valued cases.

One important property of linear operators, which is a natural requirement in systems theory, is *causality,* namely, the condition that if $u(t) = 0$ for all $t < t_0$, then $(Tu)(t) = 0$ for all $t < t_0$. We may summarise this as "we cannot influence the past", or "it's all water under the bridge now". Note that if A is a bounded shift-invariant operator defined on the whole of $\ell^2(\mathbb{Z}_+)$, then $A(R^{t_0}u) = R^{t_0}(Au)$, which easily implies the causality condition. Similar results hold on $L^2(0, \infty)$, assuming now that A commutes with all right shifts R_λ for $\lambda \geq 0$. However, it is easy to see that the left shift $R^* : \ell^2(\mathbb{Z}) \to \ell^2(\mathbb{Z})$ is not causal and to write down a similar example on $L^2(\mathbb{R})$. We study causality in more detail in Section 3.3.

When we use the term *linear system,* we shall in general mean a closed, causal, shift-invariant operator defined on one of the above spaces. We begin by studying bounded operators, for which the results we want can be derived directly in a very simple way.

Theorem 3.2.1 *Let* $T : H^2(\mathbb{D}, \mathbb{C}^m) \to H^2(\mathbb{D}, \mathbb{C}^p)$ *be a bounded linear operator that commutes with the right shift operator* S. *Then there is a function* $G \in H^\infty(\mathbb{D}, \mathcal{L}(\mathbb{C}^m, \mathbb{C}^p))$ *such that we have*

$$(Tu)(z) = G(z)u(z) \qquad \text{for all } u \in H^2(\mathbb{D}, \mathbb{C}^m).$$

Moreover $\|T\| = \|G\|_\infty$.

Proof: We begin with the scalar case, $m = p = 1$. Let $G(z) = (Te_0)(z)$, where e_0 is the function constantly equal to 1. Clearly G is an analytic function of z. If we let e_n denote the function $S^n e_0$ for $n = 0, 1, \ldots$, then, since $TS^n = S^nT$, we have $(Te_n)(z) = (S^nTe_0)(z) = z^nG(z) = G(z)e_n(z)$. Now the linear span of the (e_n) is dense in H^2, and so it follows by continuity that $(Tu)(z) = G(z)u(z)$ for all $u \in H^2$.

For each $w \in \mathbb{D}$ let $k_w \in H^2$ denote the reproducing kernel $z \mapsto (1 - \overline{w}z)^{-1}$, so that $\langle f, k_w \rangle = f(w)$ for all $f \in H^2$. Now

$$\langle u, T^*k_w \rangle = \langle Tu, k_w \rangle = G(w)u(w) = \langle u, \overline{G(w)}k_w \rangle$$

for all $u \in H^2$, and hence $T^*k_w = \overline{G(w)}k_w$ for all $w \in \mathbb{D}$. Hence $|G(w)| \leq \|T^*\| = \|T\|$, and so $\|G\|_\infty \leq \|T\|$. We already have $(Tu)(w) = G(w)u(w)$, and so we can deduce that $\|Tu\| \leq \|G\|_\infty\|u\|$, which gives $\|T\| = \|G\|_\infty$.

We may reduce the vector-valued case to the scalar case as follows. By looking at individual components of the vectors, we may regard $H^2(\mathbb{D}, \mathbb{C}^m)$ as $(H^2(\mathbb{D}))^m$, and similarly for $H^2(\mathbb{D}, \mathbb{C}^p)$, in which case $T = (T_{ij})_{i=1\ j=1}^{p\ \ m}$ is given as a $p \times m$ matrix of shift-invariant operators and thus corresponds to an $H^\infty(\mathbb{D}, \mathbb{C}^{p\times m})$-valued function. That is, we have $(Tu)(z) = G(z)u(z)$, where $u(z)$ and $(Tu)(z)$ are vectors and $G(z)$ is a matrix, for each $z \in \mathbb{D}$. We clearly have $\|T\| \leq \|G\|_\infty$, and it remains to show the converse inequality.

For each $v \in \mathbb{C}^p$, consider the function

$$k_w \otimes v : z \mapsto (1 - \overline{w}z)^{-1}v.$$

Now, for $f \in H^2(\mathbb{D}, \mathbb{C}^m)$ we have $\langle f, T^*(k_w \otimes v) \rangle = \langle G(w)f(w), v \rangle$, and so

$$T^*(k_w \otimes v) = k_w \otimes G(w)^*v.$$

Letting w vary in $\mathbb{D}$ and v on the unit sphere of $\mathbb{C}^p$, we deduce finally that $\|T\| = \|T^*\| \geq \|G\|_\infty$, and so we have equality. $\qquad\square$

The function G in the above theorem, as well as the ones that follow, is commonly called a *transfer function*.

Corollary 3.2.2 *Any bounded operator T on $\ell^2(\mathbb{Z}_+)$ that commutes with the right shift R has the form of a convolution operator $y = Tu$, where*

$$(Tu)(k) = \sum_{j=0}^{k} h(j)u(k-j).$$

Moreover, $h(0), h(1), \dots$ are the Fourier coefficients of an H^∞ transfer function G with $\|G\|_\infty = \|T\|$.

Proof: This is a direct translation of Theorem 3.2.1, using the standard unitary correspondence between $\ell^2(\mathbb{Z}_+)$ and H^2, which associates $(a_n)_0^\infty$ with $\sum_{n=0}^\infty a_n z^n$. $\square$

Analogous results hold for $H^2(\mathbb{C}_+)$, as follows. Recall that for $\lambda \geq 0$, the shift operator S_λ is defined on a (possibly vector-valued) Hardy space by the formula $(S_\lambda G)(s) = e^{-\lambda s}G(s)$.

Theorem 3.2.3 *Let $T : H^2(\mathbb{C}_+, \mathbb{C}^m) \to H^2(\mathbb{C}_+, \mathbb{C}^p)$ be a bounded linear operator that commutes with all shift operators S_λ for $\lambda \geq 0$. Then there is a function $G \in H^\infty(\mathbb{C}_+, \mathcal{L}(\mathbb{C}^m, \mathbb{C}^p))$ such that we have*

$$(Tu)(s) = G(s)u(s) \qquad \text{for all } u \in H^2(\mathbb{C}_+, \mathbb{C}^m).$$

Moreover $\|T\| = \|G\|_\infty$.

Proof: We consider only the scalar case and leave the vectorial case as an exercise. Let $\hat{e}$ denote the $H^2(\mathbb{C}_+)$ function $\hat{e}(s) = \frac{1}{s+1}$, which is the Laplace transform of the function $e : t \mapsto e^{-t}$ in $L^2(0, \infty)$. Note that the closed linear span of all the translates $R_\lambda e$ is dense in $L^2(0, \infty)$, since if $g \in L^2(0, \infty)$ and

$\langle g, R_\lambda e \rangle = 0$ for all $\lambda \geq 0$, then $e^\lambda \int_\lambda^\infty g(t)e^{-t} dt = 0$ for all $\lambda \geq 0$, and so $g = 0$.

Let $G(s) = (T\hat{e})(s)/\hat{e}(s)$ for each $s \in \mathbb{C}_+$. Clearly G is analytic in $\mathbb{C}_+$ and we see also that

$$T(S_\lambda \hat{e})(s) = S_\lambda T\hat{e}_s = e^{-\lambda s}G(s)\hat{e}(s) = G(s)(S_\lambda \hat{e})(s)$$

for each $\lambda \geq 0$. A similar formula is now true for any $u \in H^2(\mathbb{C}_+)$, since u is in the closed linear span of the $S_\lambda \hat{e}$; explicitly, we can find a sequence $u_n \to u$ with $Tu_n(s) = G(s)u_n(s)$, from which we have $(Tu)(s) = G(s)u(s)$, since the evaluation mapping is continuous.

As in the proof of Theorem 3.2.1, we see that the reproducing kernel functions k_s are eigenvectors of T^*, since

$$\langle u, T^*k_s \rangle = G(s)u(s) = \langle u, \overline{G(s)}k_s \rangle$$

for every $s \in \mathbb{C}_+$ and $u \in H^2(\mathbb{C}_+)$. Thus $\|G\|_\infty \leq \|T\|$; again, $(Tu)(s) = G(s)u(s)$, so $\|Tu\|_2 \leq \|G\|_\infty \|u\|_2$, and so we have $\|T\| = \|G\|_\infty$. $\square$

To transfer this back to a characterization of shift-invariant operators on $L^2(0, \infty)$ is not entirely straightforward. The problem is that it is hard to describe the inverse Laplace transform of a general $H^\infty(\mathbb{C}_+)$ function. The following class is often used in practice. Suppose that $g \in L^1(0, \infty)$, that $0 \leq \tau_1 \leq \tau_2 \leq \ldots$, and that $(h_j)_1^\infty$ are scalars with $\sum_{j=1}^\infty |h_j| < \infty$. Then we can define a convolution operator T_h associated with the *impulse response* h, and expressed by the formula $h(t) = g(t) + \sum_{j=1}^\infty h_j \delta(t - \tau_j)$, by

$$(T_h u)(t) = \int_{r=0}^t g(r)u(t - r)\, dr + \sum_{\{j:\tau_j \leq t\}} h_j u(t - t_j). \tag{3.3}$$

Taking Laplace transforms, we see that $(LT_h u)(s) = G(s)(Lu)(s)$ for $s \in \mathbb{C}_+$, where

$$G(s) = (Lg)(s) + \sum_{j=1}^\infty a_j e^{-\tau_j s},$$

and $G \in H^\infty(\mathbb{C}_+)$. This idea will recur later when we discuss delay systems in Chapter 6.

One fairly immediate consequence of the previous results is a characterization of the bounded causal shift-invariant operators on $\ell^2(\mathbb{Z})$ and $L^2(\mathbb{R})$.

Corollary 3.2.4 *Let T be a causal bounded shift-invariant operator on $\ell^2(\mathbb{Z})$ or $L^2(\mathbb{R})$, corresponding in the usual way to a shift-invariant operator T' on $L^2(\mathbb{T})$, respectively $L^2(i\mathbb{R})$. Then there is a function $G \in H^\infty(\mathbb{D})$, respectively $H^\infty(\mathbb{C}_+)$, such that $(T'u)(z) = G(z)u(z)$ for all $u \in L^2(\mathbb{T})$, respectively $L^2(i\mathbb{R})$, and z in the appropriate domain. Moreover, $\|T\| = \|T'\| = \|G\|_\infty$.*

Proof: The result for $\ell^2(\mathbb{Z})$ can be seen as follows. Since T is causal, the operator T has $\ell^2(\mathbb{Z}_+)$ as an invariant subspace, and thus T' has $H^2(\mathbb{D})$ as an invariant subspace. Now by Theorem 3.2.1 we have the formula $(T'u)(z) = G(z)u(z)$ for all $u \in H^2(\mathbb{D})$; but the shift-invariance of T' implies the same formula on $S^{-n}H^2(\mathbb{D})$ for any n, since $T'S^n = S^nT'$. Now $\bigcup_{n=0}^{\infty} S^{-n}H^2(\mathbb{D})$ is dense in $L^2(\mathbb{T})$, and so we deduce the result for the whole of $L^2(\mathbb{T})$.

The result for $L^2(\mathbb{R})$ follows similarly, using Theorem 3.2.3. $\square$

Now we wish to look at a more general situation, when a shift-invariant operator is no longer bounded, and it is here that our work on graphs and subspaces bears fruit.

The following theorem, given in [42], characterizes the closed operators with shift-invariant graphs on H^2. We give the full vector-valued case, but we do not specify whether we are discussing $H^2(\mathbb{D})$ or $H^2(\mathbb{C}_+)$, as there is no difference.

Theorem 3.2.5 (Georgiou–Smith) *Let $T : \mathcal{D}(T) \to H^2(\mathbb{C}^p)$ be a closed shift-invariant operator with $\mathcal{D}(T) \subseteq H^2(\mathbb{C}^m)$. Then there exist $r \leq m$, a nonsingular function $M \in H^{\infty}(\mathcal{L}(\mathbb{C}^r, \mathbb{C}^m))$, and $N \in H^{\infty}(\mathcal{L}(\mathbb{C}^r, \mathbb{C}^p))$ such that*

$$\mathcal{G}(T) = \begin{pmatrix} M \\ N \end{pmatrix} H^2(\mathbb{C}^r) = \Theta H^2(\mathbb{C}^r), \tag{3.4}$$

where $\Theta = \begin{pmatrix} M \\ N \end{pmatrix}$ is inner, that is, $\|\Theta u\| = \|u\|$ for all $u \in H^2(\mathbb{C}^r)$.

Proof: Since the graph $\mathcal{G}(T)$ is a closed shift-invariant subspace of $H^2(\mathbb{C}^{m+p})$, the vectorial form of the Beurling–Lax theorem above, Theorem 3.1.7, shows that $\mathcal{G}(T)$ has the required form. The condition that M is non-singular (and hence $r \leq m$) is there because we need $Mw = 0$ to imply that $w = 0$ in order that $\mathcal{G}(T)$ be a graph. $\square$

We also mention the analogous result for closed shift-invariant systems on L^2.

Theorem 3.2.6 *Let $T : \mathcal{D}(T) \to L^2(\mathbb{T}, \mathbb{C}^p)$ be a closed shift-invariant operator with $\mathcal{D}(T) \subseteq L^2(\mathbb{T}, \mathbb{C}^m)$. Then there exists a measurable projection-valued function $\Theta : \mathbb{T} \to L(\mathbb{C}^{m+p})$ such that*

$$\mathcal{G}(T) = \Theta L^2(\mathbb{T}, \mathbb{C}^{m+p}) = \begin{pmatrix} M \\ N \end{pmatrix} L^2(\mathbb{T}, \mathbb{C}^{m+p}), \quad say.$$

Moreover, if $Mw = 0$ for some $w \in L^2(\mathbb{C}^{m+p})$, then $Nw = 0$ also.

This follows immediately from Wiener's theorem 3.1.6, the last condition being there to guarantee that $\mathcal{G}(T)$ is indeed a graph.

In the case $m = p = 1$, this can be expressed more transparently as follows; an analogous expression for the multivariable case, which we do not need, is given in [58].

Theorem 3.2.7 Let $T : \mathcal{D}(T) \to L^2(\mathbb{T})$ be a closed shift-invariant operator with $\mathcal{D}(T) \subseteq L^2(\mathbb{T})$. Then we have

$$\mathcal{G}(T) = GL^2(\mathbb{T}),$$

where $G = \begin{pmatrix} g_1 \\ g_2 \end{pmatrix} \in L^\infty(\mathbb{T}, \mathbb{C}^2)$ with $|G(e^{i\omega})|^2 \in \{0, 1\}$ a.e. and $|g_2(e^{i\omega})| \neq 1$ a.e.

Proof: As in Theorem 3.2.6, $\mathcal{G}(T) = \Theta L^2(\mathbb{T}, \mathbb{C}^2)$, for some projection-valued Θ. Let $J(e^{i\omega})$ denote the image of $\Theta(e^{i\omega})$. Then, since $\mathcal{G}(T)$ is a graph, we have $J(e^{i\omega}) \neq \mathbb{C}^2$ a.e., so that for almost all ω we see that $J(e^{i\omega})$ is either the zero subspace or one-dimensional. In the case when it is zero, we take $G(e^{i\omega}) = \begin{pmatrix} 0 \\ 0 \end{pmatrix}$; otherwise we may select a unit vector $G(e^{i\omega}) = \begin{pmatrix} g_1(e^{i\omega}) \\ g_2(e^{i\omega}) \end{pmatrix}$ with $0 \leq g_1(e^{i\omega}) \leq 1$. The projection $\Theta(e^{i\omega})$ is given by

$$\Theta(e^{i\omega})v = \langle v, \begin{pmatrix} g_1(e^{i\omega}) \\ g_2(e^{i\omega}) \end{pmatrix} \rangle \begin{pmatrix} g_1(e^{i\omega}) \\ g_2(e^{i\omega}) \end{pmatrix}, \qquad (v \in \mathbb{C}^2). \tag{3.5}$$

It remains to prove that g_1 and g_2 are measurable functions on $\mathbb{T}$. Using the fact that $v^T \Theta w$ is a measurable function for every $v, w \in \mathbb{C}^2$ and (3.5), it is easy to see that g_1^2 and $g_1 g_2$ are measurable. Since $g_1(e^{i\omega}) \in [0, 1]$, it now follows that g_1 and α, given by

$$\alpha(e^{i\omega}) = \begin{cases} g_1(e^{i\omega})^{-1} & \text{if } g_1(e^{i\omega}) > 0, \\ 0 & \text{otherwise,} \end{cases}$$

are measurable. Thus $G \in L^\infty(\mathbb{T}, \mathbb{C}^2)$, which concludes the proof. $\qquad\square$

3.3 Causality

Systems theorists normally take their signal spaces to be based on either $(0, \infty)$ or $\mathbb{Z}_+$; if one does not, then certain paradoxes arise. In this section we present various examples of troublesome behaviour. We shall give our examples in discrete time (i.e., on $\mathbb{Z}$ or $\mathbb{Z}_+$), but the continuous-time case (i.e., $\mathbb{R}$ or $(0, \infty)$) is similar.

Example 3.3.1 *Writing $z = e^{i\omega}$, take $\mathcal{D}(T)$ to be the subspace of $L^2(\mathbb{T})$ consisting of all functions u represented as*

$$u(z) = p(z, 1/z) + q(z, 1/z)\exp(1/z),$$

with p, q polynomials in z and $1/z$. This is dense in $L^2(\mathbb{T})$. Define the operator T on $\mathcal{D}(T)$ by $(Tu)(z) = p(z, 1/z)$.

In the above example, T is causal, since if $u(z) = \sum_{n=N}^{\infty} \hat{u}(n)z^n$, then it is easily seen that the q term must be zero. Hence $Tu = u$ for such functions. Now T is a causal linear shift-invariant operator, but clearly T has no convolution representation of the form

$$(\widehat{Tu})(n) = \sum_{k=0}^{\infty} h(k)\hat{u}(n-k), \qquad \text{for all } n \in \mathbb{Z}, \tag{3.6}$$

since T is the identity on all finitely-supported sequences. It can be shown directly that this operator is not closable (see Exercise 17).

We now give a characterization of the graphs of causal shift-invariant systems defined on H^2. The vectorial case is more complicated, requiring some ring-theoretic technicalities, and can be found in [42], where the result is expressed in terms of the greatest common divisors of the minors of the analytic matrix-valued functions involved.

Theorem 3.3.2 *Let $T : \mathcal{D}(T) \to H^2(\mathbb{D})$ be a closed operator with $\mathcal{D}(T) \subseteq H^2(\mathbb{D})$ and graph $\mathcal{G}(T) = \begin{pmatrix} M \\ N \end{pmatrix} H^2(\mathbb{D})$, where M and N lie in H^∞, as in Theorem 3.2.5. Then T is causal if and only if, whenever z^m divides $M(z)$ in $H^\infty(\mathbb{D})$ for some $m > 0$, then z^m also divides $N(z)$. For operators T on $H^2(\mathbb{C}_+)$, with the same notation, T is causal if and only if, whenever $e^{-s\tau}$ divides $M(s)$ in $H^\infty(\mathbb{C}_+)$ for some $\tau > 0$, then $e^{-s\tau}$ also divides $N(s)$.*

Proof: Since the function $\begin{pmatrix} M \\ N \end{pmatrix}$ lies in the graph of T, it is clear that the divisibility condition is necessary.

The condition is also sufficient, since if m is the maximal integer such that z^m divides $M(z)$ in H^∞, and z^r divides $M(z)u(z)$ for some $r \geq m$, then necessarily z^{r-m} divides $u(z)$ and so z^r divides $N(z)u(z)$: this is precisely what we need to establish causality.

The proof for the continuous-time case, where we work on $\mathbb{C}_+$, is similar, and we omit it. $\qquad\square$

We now turn our attention back to $L^2(\mathbb{T})$. When closed shift-invariant operators have graphs of the form $\begin{pmatrix} g_1 \\ g_2 \end{pmatrix} L^2(\mathbb{T})$, as in Theorem 3.2.7, then it is natural to think of g_2/g_1 as the transfer function (so that $u = g_1 h$ is mapped to $y = g_2 h$), but this may only make sense on the unit circle, however.

Here is a characterization of causality for closed shift-invariant systems on $\ell^2(\mathbb{Z})$, which as usual we reformulate using the equivalence with $L^2(\mathbb{T})$. To characterize causality, we may assume without loss of generality that $\mathcal{D}(T)$ actually contains some functions u with $\hat{u}(k) = 0$ for $k < 0$. That is, we assume that $\mathcal{D}(T) \cap H^2(\mathbb{D}) \neq \{0\}$.

We need to introduce the *Smirnoff class* $\mathcal{N}^+$, consisting of those analytic functions $f : \mathbb{D} \to \mathbb{C}$ that can be written as $f = f_1/f_2$ with f_1, $f_2 \in H^\infty$ and f_2 outer. This is bigger than H^∞, for example, the function $z \mapsto 1/(z-1)$ is in the class.

Theorem 3.3.3 *Let T be a closed shift-invariant system on $L^2(\mathbb{T})$ with graph* $\mathcal{G}(T) = \begin{pmatrix} g_1 \\ g_2 \end{pmatrix} L^2(\mathbb{T})$, *as in Theorem 3.2.7. Suppose that $\mathcal{D}(T) \cap H^2(\mathbb{D}) \neq \{0\}$. Then T is causal if and only if g_2/g_1 lies in the Smirnoff class $\mathcal{N}^+$.*

Proof: Take an arbitrary non-zero $u \in \mathcal{D}(T) \cap H^2(\mathbb{D})$, and let $v \in L^2(\mathbb{T})$ satisfy $u = g_1 v$ and $Tu = g_2 v$. If $g_2/g_1 \in \mathcal{N}^+$, then $v = (g_2/g_1)u$ is analytic in the disc, and hence it lies in $H^2(\mathbb{D})$. Thus T is causal.

For the converse, let θ be the inner factor of u, and let

$$\begin{pmatrix} h_1 \\ h_2 \end{pmatrix} = \begin{pmatrix} g_1 \\ g_2 \end{pmatrix} v/\theta \in \mathcal{G}(T).$$

Now $h_1 = u/\theta \in H^2(\mathbb{D})$ and is outer, and $h_2 = Th_1 \in H^2(\mathbb{D})$. Hence $h_2/h_1 \in \mathcal{N}^+$; but this is the same as g_2/g_1. $\qquad\square$

Example 3.3.1 above alerted us to the fact that not all causal systems on $\ell^2(\mathbb{Z})$ are closable. Fortunately, the situation is more satisfactory for convolution systems, as the following result shows. In this case we do not even require causality.

Theorem 3.3.4 *Let $T : \mathcal{D}(T) \to \ell^2(\mathbb{Z})$ be a convolution system, defined on $\mathcal{D}(T) \subseteq \ell^2(\mathbb{Z})$ by the convolution equation*

$$(Tu)(t) = \sum_{k=-\infty}^{\infty} g(k)u(t-k), \qquad \text{for } u \in \mathcal{D}(T).$$

If $\mathcal{D}(T)$ is dense, then T is closable. In particular, if $\mathcal{D}(T) \cap \ell^2(\mathbb{Z}_+) \neq \{0\}$, then T is closable.

Proof: We see that an adjoint of T, namely $T^* : \mathcal{D}(T^*) \to \ell^2(\mathbb{Z})$, is given by

$$(T^*u)(t) = \sum_{m=-\infty}^{\infty} \overline{g(-m)}u(\tau - m),$$

and its domain includes all $u \in \ell^2(\mathbb{Z})$ such that the sequence $(\overline{u(-n)})$ lies in $\mathcal{D}(T)$, which is dense. Since T^* has a dense domain, it follows by Remark 2.1.7 that T is closable.

It remains to show that $\mathcal{D}(T)$ is dense whenever it contains a non-zero vector $u \in L^2(\mathbb{Z}_+)$. Now it is convenient to work in the frequency domain. We shall show that the linear span of the functions $(S^n h)_{n \in \mathbb{Z}}$ in $L^2(\mathbb{T})$ is dense whenever $h \in H^2(\mathbb{D}) \setminus \{0\}$. So suppose that $g \in L^2(\mathbb{T})$ and $\langle g, S^n h \rangle = 0$ for all $n \in \mathbb{Z}$. Thus

$$\frac{1}{2\pi} \int_0^{2\pi} g(e^{i\omega})e^{-in\omega}h(e^{i\omega})\,d\omega = 0 \qquad \text{for all } n \in \mathbb{Z}.$$

We deduce that $g \cdot h$, the pointwise product, is zero almost everywhere. Since $h \in H^2$, we know that $h \neq 0$ a.e. on $\mathbb{T}$ by Theorem 1.2.3; hence $g = 0$ a.e. This establishes the result.   □

Example 3.3.5 *It is possible to define a shift-invariant (necessarily causal) operator with non-trivial domain in $\ell^2(\mathbb{Z})$ but trivial domain in $\ell^2(\mathbb{Z}_+)$, although physically this seems rather implausible.*[1] *Let*

$$G(z) = \exp(-((1 - z)/(1 + z))^2).$$

Then it can be verified that $G = 1/G_0$, where $G_0 \in L^\infty(\mathbb{T})$, but that it is not possible to write $G = g_2/g_1$ with $g_1 \in H^2(\mathbb{D})$ and $g_2 \in L^2(\mathbb{T})$.

We conclude this section with a discussion of the Georgiou–Smith paradox of 1995 [43], which showed that the closure of a simple convolution system need no longer be causal. We present a discrete-time version of their very instructive example. As usual, $(e_n)_{-\infty}^{\infty}$ is the standard orthonormal basis of $\ell^2(\mathbb{Z})$.

Example 3.3.6 *Define the system $T : \mathcal{D}(T) \to \ell^2(\mathbb{Z})$ with $\mathcal{D}(T) \subset \ell^2(\mathbb{Z})$ by*

$$\begin{aligned}
(Tu)(t) &= \sum_{n=0}^{\infty} 2^n u(t - n) \\
&= u(t) + 2u(t - 1) + 4u(t - 2) + \ldots.
\end{aligned}$$

We take $\mathcal{D}(P) = \{u \in \ell^2(\mathbb{Z}) : Tu \in \ell^2(\mathbb{Z})\}$.

The operator T is causal, and it will be closable, by Theorem 3.3.4, since the vector $u = e_0 - 2e_1 \in \mathcal{D}(T)$. Indeed, we have $Tu = e_0$. However, the closure of the system is no longer causal.[2]

---

[1] It seems to correspond to a scientific experiment that can only take place if it is already taking place.

[2] We can build an approximate time machine, even if we cannot build a genuine one.

To see this, let $u_n = 2^{-n}e_{-n} - e_0$ for $n \in \mathbb{N}$. Now

$$Tu_n = \sum_{j=-n}^{-1} 2^j e_j.$$

In the limit, as $n \to \infty$, we have a non-causal extension, since

$$u_n \to -e_0 \qquad \text{and} \qquad Tu_n \to \sum_{j=-\infty}^{-1} 2^j e_j.$$

We know from Theorem 3.3.3 why this paradox can occur. The transfer function g_2/g_1 does not lie in the Smirnoff class. Indeed, it is not hard to see from the previous results that a causal convolution system of the form (3.6) on $\ell^2(\mathbb{Z})$ whose domain contains non-trivial sequences in $\ell^2(\mathbb{Z}_+)$ has causal closure and only if the function

$$G(z) = \sum_{n=0}^{\infty} g(n)z^n \qquad \text{lies in } \mathcal{N}^+.$$

In our example, $G(z) = 1/(1-2z)$, which is not in $\mathcal{N}^+$.

Another superficially similar example, which does indeed have causal closure, is given by

$$(Tu)(t) = \sum_{n=0}^{\infty}(n+1)^2 u(t-n),$$

where now $G(z) = 1/(1-z)^2$, which lies in $\mathcal{N}^+$.

Note that defining transfer functions by power series begins to appear a little misleading here. The function $G(z) = 1/(1-2z)$ has two power series expansions, namely,

$$G(z) = \begin{cases} \sum_{k=0}^{\infty} 2^k z^k & \text{if } |z| < 1/2, \\ -\sum_{k=1}^{\infty} z^{-k}/2^k & \text{if } |z| > 1/2. \end{cases}$$

These are different aspects of the same system, even though one is apparently causal and the other is not.

This last example was included, not in order to confuse the reader, but at least to highlight some pitfalls in this subject. We shall not need to lose further sleep on this.

3.4 The commutant lifting theorem

The theorem that we shall now present, although stated in the language of abstract Hilbert space operators, has many applications in approximation and interpolation problems in Hardy classes. We shall need it in a systems theory context in Chapter 4, but for the moment we shall look at some of its other corollaries.

Definition 3.4.1 *Let* $T : \mathcal{H} \to \mathcal{H}$ *be an operator. A lifting of* T *is an operator* $U : \mathcal{H}' \to \mathcal{H}'$, *defined on a Hilbert space* $\mathcal{H}' \supseteq \mathcal{H}$, *such that* $P_\mathcal{H} U = T P_\mathcal{H}$; *here* $P_\mathcal{H}$ *is the orthogonal projection from* $\mathcal{H}'$ *onto* $\mathcal{H}$. *Thus, in matrix notation* $U = \begin{pmatrix} T & 0 \\ X & Y \end{pmatrix}$ *acting on* $\mathcal{H}' = \mathcal{H} \oplus (\mathcal{H}' \ominus \mathcal{H})$.

Note that for $k \geq 1$, one has $U^k = \begin{pmatrix} T^k & 0 \\ X_k & Y^k \end{pmatrix}$ for some operator X_k, and thus U^k is a lifting of T^k.

If T is a contraction, then it possesses an isometric lifting U (see Exercise 21); for example, the unilateral left shift $S^* : H^2 \to H^2$ has the bilateral left shift $U^* : L^2(\mathbb{T}) \to L^2(\mathbb{T})$ as an isometric (even unitary) lifting.

The following result is one powerful form of the commutant lifting theorem.

Theorem 3.4.2 *Let* $T : \mathcal{H} \to \mathcal{H}$ *be a contraction and* $V : \mathcal{K} \to \mathcal{K}$ *an isometry. Let* $A : \mathcal{K} \to \mathcal{H}$ *be a bounded operator such that* $TA = AV$. *Let* $U : \mathcal{H}' \to \mathcal{H}'$ *be an isometric lifting of* T. *Then there is an operator* $B : \mathcal{K} \to \mathcal{H}'$ *such that*

1. $A = P_\mathcal{H} B$;

2. $\|A\| = \|B\|$; *and*

3. $UB = BV$.

Proof:　We may assume without loss of generality that $\|A\| = 1$. Write $\mathcal{L} = \overline{(U - T)\mathcal{H}} = \overline{X\mathcal{H}} \subseteq \mathcal{H}' \ominus \mathcal{H}$. The first step is to consider operators B of the form

$$B = A + \sum_{k=0}^{\infty} U^k B_k$$

with $B_k : \mathcal{K} \to \mathcal{L}$ and the relations $B_k V = B_{k-1}$ for $k \geq 0$, where $B_{-1} = (U-T)A$. Note that Condition 1 follows because $U^k B_k$ maps into $\mathcal{H}' \ominus \mathcal{H}$. Also, Condition 3 (assuming convergence of the sum, which we shall establish later) follows because

$$
\begin{aligned}
UB - BV &= UA + \sum_{k=0}^{\infty} U^{k+1} B_k - \left(AV + \sum_{k=0}^{\infty} U^k B_k V \right) \\
&= (TA - AV) + B_{-1} + \sum_{k=0}^{\infty} (U^{k+1} B_k - U^k B_{k-1}) \\
&= 0.
\end{aligned}
$$

The convergence of the expression for B and the norm condition on B will follow by an inductive choice of (B_k). Note that, for $k > m$,

$$\langle U^k B_k x, U^m B_m x \rangle = \langle U^{k-m} B_k x, B_m x \rangle = 0,$$

since, for $y, z \in \mathcal{H}$ and $r \geq 1$, we have

$$
\begin{aligned}
\langle U^r (U - T)y, (U - T)z \rangle &= \langle U^r y, z \rangle - \langle U^{r-1} Ty, z \rangle - \langle U^{r+1} y, Tz \rangle + \langle U^r Ty, Tz \rangle \\
&= \langle U^r y, z \rangle - \langle T^{r-1} Ty, z \rangle - \langle T^{r+1} y, Tz \rangle + \langle T^r Ty, Tz \rangle \\
&= 0,
\end{aligned}
$$

and also $\langle Ay, U^k B_k z \rangle = 0$ for $k \geq 0$. Hence

$$
\|Bx\|^2 = \|Ax\|^2 + \sum_{k=0}^{\infty} \|B_k x\|^2. \tag{3.7}
$$

Suppose now that for some $N \geq 0$ we have constructed $B_{-1}, \ldots, B_{N-1}$ satisfying the algebraic relations $B_k V = B_{k-1}$ for $0 \leq k \leq N - 1$ and $B_{-1} = (U - T)A$, such that

$$
s_N(x) = \|Ax\|^2 + \sum_{k=0}^{N-1} \|B_k x\|^2 \leq \|x\|^2 \qquad \text{for all} \quad x \in \mathcal{K}.
$$

(This is clearly possible for $N = 0$.) The operator $R_N = I - A^* A - \sum_{k=0}^{N-1} B_k^* B_k$ is therefore positive (i.e., $\langle R_N x, x \rangle \geq 0$ for all $x \in \mathcal{K}$), and so we let $D_N : \mathcal{K} \to \mathcal{K}$ denote its positive square root – see Exercise 20 for a proof of the existence of such a square root. Note that $D_0^2 = R_0 = I - A^* A$.

Now

$$
s_N(Vx) = \|AVx\|^2 + \sum_{k=0}^{N-1} \|B_k Vx\|^2 = \|AVx\|^2 + \|B_{-1}x\|^2 + \sum_{k=0}^{N-2} \|B_k x\|^2,
$$

since $B_k V = B_{k-1}$, and also

$$
\|B_{-1}x\|^2 + \|AVx\|^2 = \|(U - T)Ax\|^2 + \|TAx\|^2 = \|Ax\|^2,
$$

since $(U - T)Ax \perp TAx$. Therefore

$$
s_N(Vx) = s_N(x) - \|B_{N-1}x\|^2 \leq \|x\|^2 - \|B_{N-1}x\|^2.
$$

Thus

$$
\|B_{N-1}x\|^2 \leq \|x\|^2 - s_N(Vx) = \|Vx\|^2 - \|AV\|^2 - \sum_{k=0}^{N-1} \|B_k Vx\|^2 = \|D_N Vx\|^2.
$$

The same inequality holds for $N = 0$, since

$$
\|B_{-1}x\|^2 \leq \|Ax\|^2 - \|AVx\|^2 \leq \|x\|^2 - \|AVx\|^2 = \|Vx\|^2 - \|AVx\|^2 = \|D_0 Vx\|^2.
$$

We therefore have the inequality $\|B_{N-1}x\| \leq \|D_N V x\|$ for all $x \in \mathcal{K}$. This implies that there is a contraction $E_N : \mathcal{K} \to \mathcal{L}$ such that $B_{N-1} = E_N D_N V$ (see Exercise 22 for more details).

Now let $B_N = E_N D_N$, so that we have $B_N V = B_{N-1}$ and

$$\|B_N x\|^2 \leq \|D_N x\|^2 = \|x\|^2 - \|Ax\|^2 - \sum_{k=0}^{N-1} \|B_k x\|^2,$$

that is, $s_{N+1}(x) \leq \|x\|^2$, for $x \in \mathcal{K}$. Thus, by induction on N we may choose (B_N) such that, on letting $N \to \infty$, we have $\|B\| \leq 1$, and this completes the proof of the theorem. □

We shall now see some applications of the commutant lifting theorem to vector-valued Hardy spaces. In fact, the analogous results hold in operator-valued Hardy spaces (appropriately defined), with almost identical proofs, at least if all the Hilbert spaces involved are separable.

Theorem 3.4.3 *Let $F \in H^\infty(\mathcal{L}(\mathbb{C}^m, \mathbb{C}^p))$ and let $\Theta \in H^\infty(\mathcal{L}(\mathbb{C}^n, \mathbb{C}^p))$ be inner (so that $\Theta(e^{i\omega})$ is an isometry for almost all $e^{i\omega} \in \mathbb{T}$). Then*

$$\inf\{\|F - \Theta G\|_\infty : G \in H^\infty(\mathcal{L}(\mathbb{C}^m, \mathbb{C}^n))\} = \|P_\mathcal{X} M_F\|, \qquad (3.8)$$

where $P_\mathcal{X}$ is the orthogonal projection from $H^2(\mathbb{C}^p)$ onto $\mathcal{X} = H^2(\mathbb{C}^p) \ominus \Theta H^2(\mathbb{C}^n)$ and $M_F : H^2(\mathbb{C}^m) \to H^2(\mathbb{C}^p)$ is the operator of multiplication by F. Moreover, the infimum is attained.

Proof: The easy part of the theorem is the fact that "$\geq$" holds in (3.8): for if $u \in H^2(\mathbb{C}^m)$, then

$$P_\mathcal{X} M_F u = P_\mathcal{X}(Fu) = P_\mathcal{X}((F - \Theta G)u)$$

for any $G \in H^\infty(\mathcal{L}(\mathbb{C}^m, \mathbb{C}^n))$, and thus $\|P_\mathcal{X} M_F\| \leq \|F - \Theta G\|_\infty$. Now take the infimum over G.

For the converse inequality, we shall apply the commutant lifting theorem with $\mathcal{H} = \mathcal{X}$, $\mathcal{H}' = H^2(\mathbb{C}^p)$ and $\mathcal{K} = H^2(\mathbb{C}^m)$. Write S for the unilateral shift ("multiplication by z") on $H^2(\mathbb{C}^m)$, $H^2(\mathbb{C}^n)$ or $H^2(\mathbb{C}^p)$, according to context. Let $A = P_\mathcal{X} M_F : H^2(\mathbb{C}^m) \to \mathcal{X}$, $T = P_\mathcal{X} S_{|\mathcal{X}} : \mathcal{X} \to \mathcal{X}$ and $V = S$, and note that

$$TA = P_\mathcal{X} S P_\mathcal{X} M_F = P_\mathcal{X} S M_F = P_\mathcal{X} M_F S = AV,$$

since if $u \in H^2(\mathbb{C}^p)$, then

$$P_\mathcal{X} S P_\mathcal{X} M_F u - P_\mathcal{X} S M_F u = P_\mathcal{X} S \Theta v$$

for some $v \in H^2(\mathbb{C}^n)$; however, $P_{\mathcal{X}} \Theta S v = 0$.

Note that T has an isometric lifting $U : H^2(\mathbb{C}^p) \to H^2(\mathbb{C}^p)$, given by $U = S$, since S maps $H^2(\mathbb{C}^p) \ominus \mathcal{X} = \Theta H^2(\mathbb{C}^n)$ into itself.

We now apply Theorem 3.4.2 to deduce the existence of an operator $B :$ $H^2(\mathbb{C}^m) \to H^2(\mathbb{C}^p)$ such that $A = P_{\mathcal{X}} B$, $\|A\| = \|B\|$ and $UB = BV$. The last identity shows that $B = M_Q$ for some $Q \in H^\infty(\mathcal{L}(\mathbb{C}^m, \mathbb{C}^p))$ with $\|Q\|_\infty = \|B\| = \|A\|$ (see, for example, Theorem 3.2.1).

Now $A = P_{\mathcal{X}} M_Q = P_{\mathcal{X}} M_F$, and so $P_{\mathcal{X}} M_{F-Q} = 0$. Thus $(F - Q)u \in \Theta H^2(\mathbb{C}^n)$ for every $u \in H^2(\mathbb{C}^m)$. It is easily verified that the mapping $u \mapsto v$ defined by $(F - Q)u = \Theta v$ is linear and bounded from $H^2(\mathbb{C}^m)$ to $H^2(\mathbb{C}^n)$. It is also shift-invariant, since $(F - Q)Su = S\Theta v = \Theta S v$, and thus $v = Gu$ for some $G \in H^\infty(\mathcal{L}(\mathbb{C}^m, \mathbb{C}^n))$.

We conclude that $F - Q = \Theta G$, that is, $F - \Theta G = Q$, with $\|Q\|_\infty = \|A\|$, as required. $\qquad \square$

We now mention a related application that links the ideas of distance and operator norm. Again, an analogous version holds in the operator-valued context. The following result is known as Nehari's theorem.

Theorem 3.4.4 *Let ϕ be a function in $L^\infty(\mathcal{L}(\mathbb{C}^m, \mathbb{C}^p))$, let $\mathcal{X}$ denote the space $L^2(\mathbb{C}^p) \ominus H^2(\mathbb{C}^p)$, and let $M_\phi : H^2(\mathbb{C}^m) \to L^2(\mathbb{C}^p)$ be the operator of multiplication by ϕ and $P_{\mathcal{X}}$ the orthogonal projection from $L^2(\mathbb{C}^p)$ into $\mathcal{X}$. Then*

$$\inf\{\|\phi - G\|_\infty : G \in H^\infty(\mathcal{L}(\mathbb{C}^m, \mathbb{C}^p))\} = \|P_{\mathcal{X}} M_\phi\| = d, \qquad say. \qquad (3.9)$$

The infimum is attained, and $P_{\mathcal{X}} M_\phi = P_{\mathcal{X}} M_\psi$ for a function $\psi \in L^\infty(\mathcal{L}(\mathbb{C}^m, \mathbb{C}^p))$ with $\|\psi\|_\infty = d$.

Proof: Note that if $\psi = \phi - G$, with $G \in H^\infty(\mathcal{L}(\mathbb{C}^m, \mathbb{C}^p))$, then

$$\|\psi\|_\infty \geq \|P_{\mathcal{X}} M_\psi\| = \|P_{\mathcal{X}} M_\phi\|,$$

since $P_{\mathcal{X}} M_G = 0$; that is, we easily obtain "$\geq$" in (3.9).

To see the converse, we shall again apply the commutant lifting theorem with $\mathcal{H} = \mathcal{X}$, $\mathcal{H}' = L^2(\mathbb{C}^p)$, and $\mathcal{K} = H^2(\mathbb{C}^m)$. Let $A = P_{\mathcal{X}} M_\phi$, and note that $TA = AV$, where V is the unilateral shift S on $H^2(\mathbb{C}^m)$ and $T = P_{\mathcal{X}} S_{|\mathcal{X}}$, the compression of the shift to $\mathcal{X}$. The (bilateral) shift operator $U : L^2(\mathbb{C}^p) \to L^2(\mathbb{C}^p)$ is an isometric lifting of T. Hence, by the commutant lifting theorem, Theorem 3.4.2, there is an operator $B : H^2(\mathbb{C}^m) \to L^2(\mathbb{C}^p)$ such that $A = P_{\mathcal{X}} B$, $UB = BV$, and $\|B\| = \|A\|$.

The operator B will clearly be an operator of multiplication by some function $\psi \in L^\infty(\mathcal{L}(\mathbb{C}^m, \mathbb{C}^p))$, and $\|\phi\|_\infty = \|B\| = \|A\|$. (To see this, one can adapt the proof of Theorem 3.2.1. We leave it as an exercise.) Moreover,

$$P_\mathcal{X} M_\phi = A = P_\mathcal{X} B = P_\mathcal{X} M_\psi,$$

so that $G = \phi - \psi$ lies in $H^\infty(\mathcal{L}(\mathbb{C}^m, \mathbb{C}^p))$. $\square$

The operator $\Gamma_\phi = P_\mathcal{X} M_\phi : H^2(\mathbb{C}^m) \to \mathcal{X}$ is known as the *Hankel operator with symbol* ϕ, and Nehari's theorem asserts that

$$\|\Gamma_\phi\| = \operatorname{dist}(\phi, H^\infty(\mathcal{L}(\mathbb{C}^m, \mathbb{C}^p))).$$

Indeed there is a minimal norm symbol $\psi \in L^\infty(\mathcal{L}(\mathbb{C}^m, \mathbb{C}^p))$ such that $\Gamma_\phi = \Gamma_\psi$ and $\|\Gamma_\psi\| = \|\psi\|_\infty$.

We shall encounter Hankel operators again rather briefly in Section 6.3, in the context of rational approximation.

Notes

There is a strong link between the study of the invariant subspace problem and the properties of operators on H^2. For example, it was shown in [93] that the problem can be reduced to a study of the invariant subspaces of the composition operator $C_\phi : H^2 \to H^2$ given by $(C_\phi f)(z) = f(\phi(z))$, where $\phi(z) = (z+2)/(1+2z)$.

The Beurling–Helson, Wiener and Beurling–Lax theorems can be found in many places; the accounts in [55, 57] are particularly elementary, but [90, 92, 115] are also recommended. The simplest proofs now available are due largely to Helson and Srinivasan [125, 126]. Lax's original work is in [72].

There is a well-developed theory of graphs of systems in the time domain, due principally to Willems. This is known as the *behavioural approach*. We are studying different questions and use different tools, but the interested reader may consult [110].

The elementary discussion of bounded shift-invariant operators draws on [105], and also the remarkable paper [142], which shows that all shift-invariant operators on $L^p(0, \infty)$ for $1 \leq p < \infty$ are represented as multiplication operators by transfer functions. The results go back to [36, 53] in the case of L^2, and the link between causality and analyticity seems to date from this time. As [142] shows by means of a complicated example, the analogous result is not valid for $L^\infty(0, \infty)$, and this corrects assertions made elsewhere in the published literature (e.g., [52]).

Theorem 3.2.7 is taken from [61].

Some attempts to discuss the graphs of linear systems in a non-Hilbertian case were made in [78], but no fully satisfactory theory has yet been developed.

Section 3.3 is based largely on [61], which in turn was much influenced by [42, 43]. Further results, especially the vectorial case, may be found in [58].

There are other problems with the full time-axis approach to linear systems, namely, since an unstable convolution operator is not closed on $\ell^2(\mathbb{Z})$, certain stabilization results, to be discussed more fully in Chapter 4, do not apply. Moreover, if we look at an interval of the form $(-\infty, T)$, to consider an experiment that has already happened, then the corresponding operators are not even closable. We refer to [76, 77] for more on this problem.

More information on the Smirnoff class can be found in [26].

The commutant lifting theorem appears first in [127] and the monograph [128]. It provides a powerful generalization of the results of Sarason [120], who gave applications to interpolation in H^∞.

The more recent literature devoted to this subject is vast, and we mention only the recent books [33, 34].

Nehari's theorem is due to Nehari [89] in the scalar case $m = p = 1$ and Page [96] in the vector case. Many books contain expositions of this result; see for example [92, 97, 107, 111, 91, 108].

Exercises

1. Let $\mathcal{H}$ be a finite-dimensional complex Banach space of dimension at least 2; by considering eigenspaces, show that every $T \in \mathcal{L}(\mathcal{H})$ has a non-trivial closed invariant subspace.

2. Give an example of a *real* finite-dimensional Hilbert space $\mathcal{H}$ and an operator $T \in \mathcal{L}(\mathcal{H})$ with no non-trivial invariant subspaces.

3. Give an example of a subspace of $L^2(\mathbb{T})$ that is invariant under S^2 but not invariant under S.

4. Show directly that no subspace of the form ϕH^2 with $|\phi| = 1$ a.e. on $\mathbb{T}$ can be 2-invariant for the shift S.

5. Show that ϕH^2 is a closed subspace of $L^2(\mathbb{T})$, whenever $|\phi| = 1$ a.e. on $\mathbb{T}$, as was asserted in the proof of Theorem 3.1.2.

6. Suppose that $f \in H^2$ and that f vanishes on a subset of $\mathbb{T}$ of positive Lebesgue measure. By considering the smallest shift-invariant subspace of H^2 containing f and using Beurling's theorem, derive a contradiction.

7. Let $\mathcal{K}$ be a closed shift-invariant subspace of $H^2(\mathbb{T}, \mathbb{C}^m)$. Show by linear algebra that $\dim \mathcal{K} \ominus S\mathcal{K} \leq m$, and give an example of a subspace of $H^2(\mathbb{T}, \mathbb{C}^2)$ for which $\dim \mathcal{K} \ominus S\mathcal{K} = 1$.

8. Suppose that $\phi \in H^2$ is an inner function and $\phi(0) = 0$. Show that the sequence $(\phi^n)_{n\geq0}$, where $\phi^0 \equiv 1$, is orthonormal, and deduce that the composition operator $C_\phi : H^2 \to H^2$, with $C_\phi f = f \circ \phi$, is an isometry.

9. Taking $\phi(z) = z^2$ in Exercise 8, show how to interpret C_ϕ as a shift operator (note that $\dim(H^2 \ominus C_\phi H^2) = \infty$).

10. Suppose that $T \in \mathcal{L}(\mathcal{H})$. Show that $\mathcal{K}$ is an invariant subspace for T if and only if $\mathcal{K}^\perp$ is an invariant subspace for T^*. Hence, find all the invariant subspaces of S^*, the adjoint of the shift on H^2.

11. Fix $a > 0$. Verify that $L^2(a, \infty)$, regarded as a closed subspace of $L^2(0, \infty)$, is invariant under all the operators $(R_\lambda)_{\lambda\geq0}$. To which inner function does it correspond in the Beurling–Lax theorem?

12. Find a description of the closed subspaces of $H^2(\mathbb{C}_+)$ invariant under the operator S_1 of multiplication by e^{-s}. In particular, find a subspace invariant under S_1 that is not invariant under all $(S_\lambda)_{\lambda\geq0}$.

13. Take $1 \leq p < \infty$. Show that any shift-invariant operator T on $\ell^p(\mathbb{Z}_+)$ has a convolution representation as in Corollary 3.2.2, where now $\|T\| \geq \|G\|_\infty$. Prove that the *exact* norm of T in the case $p = 1$ is $\sum_{k=0}^\infty |h(k)|$.

14. Let $B : \ell^\infty(\mathbb{Z}) \to \mathbb{C}$ denote a generalized limit (Banach limit), sometimes written $\mathrm{Blim}_{n \to -\infty} u(n)$, which agrees with the usual notion of limit when it exists and is shift-invariant (for the existence of such limits see [9]). Let $T : \ell^\infty(\mathbb{Z}) \to \ell^\infty(\mathbb{Z})$ be defined by $(Tu)(k) = Bu$ for all $k \in \mathbb{Z}$. Show that T is a causal shift-invariant operator on $\ell^\infty(\mathbb{Z})$ such that $Te_n = 0$ for all n (where e_n has its usual meaning). In particular, T cannot be represented as a convolution.

15. Fill in the details of the proof of the vectorial case of Theorem 3.2.3.

16. For the convolution equation (3.3) verify the relation

$$(LT_h u)(s) = G(s)(Lu)(s)$$

and the formula for $G(s)$.

17. Consider the operator T in Example 3.3.1. By finding an explicit sequence u_n converging to 0 such that $Tu_n \to y$ with $y \neq 0$, show that T is not closable.

18. Determine which of the following functions of z, analytic in $\mathbb{D}$, lie in the Smirnoff class: $1/(z^2 - 1)$, $\exp((1 - z)/(1 + z))$, $(1 - az)/(1 + az)$ for various real values of a.

19. Define an operator T on H^2 with $\mathcal{D}(T)$ equal to the set of all functions expressible as $p(z) + q(z) \exp(z)$, with p and q polynomials, by

$$T(p(z) + q(z) \exp(z)) = q(z) \exp(z).$$

Show that T is a linear shift-invariant operator that cannot be represented by a convolution on $\ell^2(\mathbb{Z}_+)$.

20. Let A be a positive self-adjoint operator with spectrum contained in $[0, M]$, say. Show that $p(A)$ is also self-adjoint for any real polynomial p. Let (p_n) be a sequence of polynomials tending uniformly to the function $t \mapsto t^{1/2}$ on

$[0, M]$ (such sequences exist by the Weierstrass approximation theorem). Show that $(p_n(A))$ is a Cauchy sequence of self-adjoint operators tending in norm to a positive self-adjoint operator B such that $B^2 = A$.

21. Let $T : \mathcal{H} \to \mathcal{H}$ be a contraction and B the square root of the positive operator $I - T^*T$, as constructed in Example 20. Show that $\|Tx\|^2 + \|Bx\|^2 = \|x\|^2$ for all $x \in \mathcal{H}$. Let $\mathcal{J} = \overline{B\mathcal{H}}$, and let $\mathcal{K} = \mathcal{H} \oplus \mathcal{J} \oplus \mathcal{J} \oplus \ldots$, the ℓ^2 direct sum. Show that the operator $U : \mathcal{K} \to \mathcal{K}$ defined by

$$U = \begin{pmatrix} T & 0 & 0 & 0 & \cdots \\ B & 0 & 0 & 0 & \cdots \\ 0 & I & 0 & 0 & \cdots \\ 0 & 0 & I & 0 & \cdots \\ 0 & 0 & 0 & I & \cdots \\ \vdots & \vdots & \vdots & \vdots & \ddots \end{pmatrix}$$

is an isometric lifting of T.

22. Suppose that $B : \mathcal{K} \to \mathcal{L}$ and $D : \mathcal{K} \to \mathcal{K}$ are Hilbert space operators such that $\|Bx\| \le \|Dx\|$ for all $x \in \mathcal{K}$. Prove that there exists a contraction $E : \mathcal{K} \to \mathcal{L}$ such that $B = ED$ (it is clear how to define E on $\mathcal{J} = \overline{D\mathcal{K}}$; define it to be 0 on $\mathcal{K} \ominus \mathcal{J}$, and show that this gives a contraction on the whole of $\mathcal{K}$).

23. Adapt the proof of Theorem 3.2.1 to show that any shift-invariant operator $B : H^2(\mathbb{C}^m) \to L^2(\mathbb{C}^p)$ is given by multiplication by a function in $L^\infty(\mathcal{L}(\mathbb{C}^m, \mathbb{C}^p))$.

24. Apply Nehari's theorem to calculate the distance of the function $z \mapsto 3z^{-1} + 2z^{-3}$ from H^∞. (The Hankel operator can be represented by a 3×3 real symmetric matrix.)

Chapter 4

Stability and stabilization

The theme of this chapter is control theory. We discuss what it means to say that a linear system is stable, and then present some of the themes of H^∞ control theory, presented from an operator-theoretic point of view.

One of the main aims of modern control theory is to achieve *robustness*, that is, the stabilization of a system subject to perturbations, measurement errors, and the like. In order to study this we require a measure of the distance between systems, and it turns out that the operator gap is the "correct" one to use. Another way of measuring distances, the so-called chordal metric between meromorphic functions, turns out to be closely related.

4.1 Stability theory

The basic signal spaces in this chapter are vector-valued $L^2(0, \infty)$ or $\ell^2(\mathbb{Z}_+)$ spaces, and we are concerned with shift-invariant input–output operators T. Our first result shows that, if the domain of such an operator is the whole space, then it is necessarily bounded (a result in *automatic continuity theory*).

Theorem 4.1.1 *Let $T : L^2(0, \infty; \mathbb{C}^m) \to L^2(0, \infty; \mathbb{C}^p)$ be an operator commuting with the right shift R_λ for some $\lambda > 0$. Then T is bounded.*

Proof: It is sufficient to prove the result for $m = p = 1$, since in general T may be represented by a $p \times m$ matrix of shift-invariant operators from $L^2(0, \infty)$ to itself.

If T is unbounded, then there is a sequence of inputs (u_k) in $L^2(0, \infty)$ such that $\|u_k\| = 1$ but $\|Tu_k\| \geq 2^k$. We shall find integers $0 = n_1 < n_2 < \dots$ such that the input $u = \sum_{k=1}^\infty R_\lambda^{n_k} u_k / 2^k$, which clearly lies in $L^2(0, \infty)$, does not produce an output Tu lying in $L^2(0, \infty)$.

To do this, we choose strictly positive integers $m_1, m_2, \ldots$ such that

$$\left(\int_{m_k}^\infty |(Tu_k)(t)|^2 \, dt \right)^{1/2} \leq 2^{-k} \qquad \text{for} \quad k = 1, 2, \ldots.$$

and define $n_j = \sum_{k<j} m_j$ (so that $n_0 = 0$). For each $j \geq 1$, we may write $u = v_j + w_j$, where $v_j = \sum_{k=1}^j R_\lambda^{n_k} u_k / 2^k$ and $w_j = \sum_{k=j+1}^\infty R_\lambda^{n_k} u_k / 2^k$. In particular, Tw_j vanishes almost everywhere on the interval $(0, n_{j+1}\lambda)$, since T commutes with R_λ, and hence $Tu = Tv_j$ a.e. on $(0, n_{j+1}\lambda)$.

Now, for $t \in (n_j\lambda, n_{j+1}\lambda)$, we have

$$(Tu)(t) = (Tv_j)(t) = \sum_{k=1}^j R_\lambda^{n_k} Tu_k(t)/2^k.$$

But for $k = j$ the $L^2(n_j\lambda, n_{j+1}\lambda)$ norm of the restriction of $R_\lambda^{n_k} Tu_k/2^k$ is at least $(2^k - 2^{-k})/2^k = 1 - 4^{-k}$, while for $k < j$ the corresponding L^2 norm is at most 4^{-k}. Hence, by the triangle inequality, the $L^2(n_j\lambda, n_{j+1}\lambda)$ norm of the restriction of Tu is at least $1 - \sum_{k=1}^j 4^{-k}$, which is greater than $2/3$.

Thus

$$\int_{n_j\lambda}^{n_{j+1}\lambda} |(Tu)(t)|^2 \, dt \geq \frac{4}{9}, \qquad \text{for} \quad j = 1, 2, \ldots,$$

which proves that $Tu \notin L^2(0, \infty)$, as asserted. $\square$

The discrete-time version of Theorem 4.1.1 is easier (see the exercises).

We say that a linear system (which from now on we take to be a closed, causal, shift-invariant operator $T : \mathcal{D}(T) \to L^2(0, \infty; \mathbb{C}^p)$ with $\mathcal{D}(T) \subseteq \mathcal{U} = L^2(0, \infty; \mathbb{C}^m)$ or, in the discrete case, $T : \mathcal{D}(T) \to \ell^2(\mathbb{Z}_+, \mathbb{C}^p)$ with $\mathcal{D}(T) \subseteq \mathcal{U} = \ell^2(\mathbb{Z}_+, \mathbb{C}^m)$) is *input–output stable* or just *stable* if $\mathcal{D}(T) = \mathcal{U}$. From the above we see that, if this happens, then T is a bounded operator. The term *gain* is sometimes used to mean $\|T\|$. Note that a bounded shift-invariant operator is unitarily equivalent to a multiplication operator (by a transfer function) between vector-valued Hardy spaces, by Theorems 3.2.1 and 3.2.3, and we shall often use this equivalent form without further comment.

Example 4.1.2 Take $\mathcal{U} = L^2(0, \infty)$ and let $T : \mathcal{D}(T) \to L^2(0, \infty)$ be defined by $(Tu)(t) = \int_0^t u(\tau) \, d\tau$. Clearly $\mathcal{D}(T) = \{u \in L^2(0, \infty) : Tu \in L^2(0, \infty)\} \neq \mathcal{U}$: for example, the function $u : t \mapsto 1/(t + 1)$ is in $L^2(0, \infty)$, whereas its integral $t \mapsto \ln(t + 1)$ is not.

It is also easy to see directly that T is unbounded on its domain, since for $M > 0$ the operator T maps $\chi_{(0,M)}(t) - \chi_{(M,2M)}(t)$, which has norm $\sqrt{2M}$, into $t\chi_{(0,M)}(t) + (2M - t)\chi_{(M,2M)}(t)$, which has L^2 norm equal to $\sqrt{2M^3/3}$.

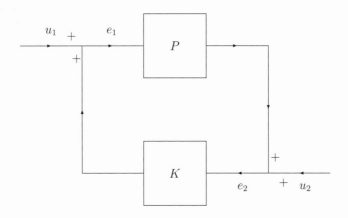

Figure 4.1: A standard feedback configuration.

Unstable systems are inconvenient from an engineering point of view, and we shall now consider the problem of stabilization. Consider the "standard feedback configuration" of Figure 4.1. By this we mean that the quantities $u_1, e_1 \in H^2(\mathbb{C}^m)$ and $u_2, e_2 \in H^2(\mathbb{C}^p)$ are related by the (in general densely defined) operators P (the plant) and K (the controller), according to the equations

$$\begin{aligned} e_1 &= Ke_2 + u_1, \\ e_2 &= Pe_1 + u_2 \end{aligned} \tag{4.1}$$

or, in matrix notation,

$$\begin{pmatrix} u_1 \\ u_2 \end{pmatrix} = \begin{pmatrix} I & -K \\ -P & I \end{pmatrix} \begin{pmatrix} e_1 \\ e_2 \end{pmatrix}. \tag{4.2}$$

The terminology that is commonly used refers to e_1 and e_2 as *errors* and u_1 and u_2 as *inputs*. Note that if $K = 0$ (i.e., no controller), then $e_1 = u_1$, and so $e_2 = Pu_1 + u_2$.

We say that the feedback system $[P, K]$ is *stable* if the mapping from (u_1, u_2) to (e_1, e_2) has a bounded extension to $H^2(\mathbb{C}^{m+p})$. Elementary algebraic calculations (see the exercises) show that, formally,

$$\begin{pmatrix} e_1 \\ e_2 \end{pmatrix} = \begin{pmatrix} (I - KP)^{-1} & K(I - PK)^{-1} \\ P(I - KP)^{-1} & (I - PK)^{-1} \end{pmatrix} \begin{pmatrix} u_1 \\ u_2 \end{pmatrix}. \tag{4.3}$$

It is clear that the stability of the "open-loop" system P is equivalent to the stability of the "closed-loop" feedback system obtained by taking $K = 0$. The plant P is *stabilizable*, if there is controller K (usually taken to be shift-invariant but not necessarily bounded) such that the feedback system $[P, K]$ given by (4.3) is stable.

For instance, continuing with Example 4.1.2, the unitarily equivalent operator P of multiplication by $1/s$, defined on a subdomain of $H^2(\mathbb{C}_+)$, can be stabilized by the negative of the identity operator, $K = -I$, on $H^2(\mathbb{C}_+)$. For the closed-loop operator is given by the multiplication operator

$$\begin{pmatrix} e_1 \\ e_2 \end{pmatrix} = M_G \begin{pmatrix} u_1 \\ u_2 \end{pmatrix}, \qquad \text{where} \quad G(s) = \begin{pmatrix} s/(s+1) & -s/(s+1) \\ 1/(s+1) & s/(s+1) \end{pmatrix},$$

and the matrix of G consists of four functions that are all in $H^\infty(\mathbb{C}_+)$.

The following observation explains why we may assume by definition that linear systems all have closed graphs.

Proposition 4.1.3 *Let $P : \mathcal{D}(P) \to H^2(\mathbb{C}^p)$ with $\mathcal{D}(P) \subset H^2(\mathbb{C}^m)$ be a linear mapping such that, for some linear $K : \mathcal{D}(K) \to H^2(\mathbb{C}^m)$ with $\mathcal{D}(K) \subset H^2(\mathbb{C}^p)$, the closed-loop operator $H(P, K) = \begin{pmatrix} (I - KP)^{-1} & K(I - PK)^{-1} \\ P(I - KP)^{-1} & (I - PK)^{-1} \end{pmatrix}$ has a bounded extension to the whole of $H^2(\mathbb{C}^{m+p})$. Then P and K are closable.*

Proof: Suppose that we have vectors v, w and a sequence (v_n) with $v_n \to v$ and $Pv_n \to w$. Then, taking $u_1 = v_n$ and $u_2 = -Pv_n$, we have the solution to (4.1) given by $e_1 = v_n$ and $e_2 = 0$. Since the mapping from (u_1, u_2) to (e_1, e_2) has a bounded extension, we see that, taking $u_1 = v$ and $u_2 = -w$, we must obtain the solution $e_1 = v$ and $e_2 = 0$. This implies that $e_2 = Pe_1 + u_2$, or $0 = Pv - w$, as required.

The proof of the fact that K is closable is similar (swap the roles of P and K).   □

Recalling Theorem 3.2.5, we know that any linear system P (i.e., causal, closed, shift-invariant) has a graph of the following form: there exist $r \leq m$, $M \in H^\infty(\mathcal{L}(\mathbb{C}^r, \mathbb{C}^m))$ nonsingular and $N \in H^\infty(\mathcal{L}(\mathbb{C}^r, \mathbb{C}^p))$ such that

$$\mathcal{G}(P) = \begin{pmatrix} M \\ N \end{pmatrix} H^2(\mathbb{C}^r) = \Theta H^2(\mathbb{C}^r), \tag{4.4}$$

where $\Theta = \begin{pmatrix} M \\ N \end{pmatrix}$ is inner, that is, $\|\Theta u\| = \|u\|$ for all $u \in H^2(\mathbb{C}^r)$. The function spaces are defined on $\mathbb{D}$ or $\mathbb{C}_+$, according to context.

To analyse the above result further, we discuss isometries (operators such that $\|Tu\| = \|u\|$ for all u) in more detail.

Lemma 4.1.4 *(i) Let $T : \mathcal{H} \to \mathcal{K}$ be a bounded linear operator between Hilbert spaces. Then T is an isometry if and only if $T^*T = I$.*

(ii) A function $G \in H^\infty(\mathbb{D}, \mathcal{L}(\mathbb{C}^r, \mathbb{C}^n))$ is inner if and only if $G(e^{i\omega})^*G(e^{i\omega}) = I$ for a.e. $\omega \in [0, 2\pi]$. An analogous result holds for $\mathbb{C}_+$.

Proof: (i) Clearly, if $T^*T = I$, then $\langle Tu, Tu \rangle = \langle T^*Tu, u \rangle = \langle u, u \rangle$ for all $u \in \mathcal{H}$, so that T is an isometry. Conversely, if T is an isometry, then we have $\langle (T^*T - I)u, u \rangle = 0$ for all u, but now $A = T^*T - I$ is Hermitian, and, by the polarization identity,

$$\langle Au, v \rangle = \frac{1}{4} \sum_{j=1}^{4} i^j \langle A(u + i^j v), u + i^j v \rangle = 0 \tag{4.5}$$

for all $u, v \in \mathcal{H}$, and hence A is the zero operator.

(ii) We see that G is inner if and only if $\langle Gu, Gv \rangle = \langle u, v \rangle$ for all $u, v \in H^2(\mathbb{C}^r)$. In the case of the disc, this can be rewritten as

$$\int_0^{2\pi} \langle G(e^{i\omega})u(e^{i\omega}), G(e^{i\omega})v(e^{i\omega}) \rangle \, d\omega = \int_0^{2\pi} \langle u(e^{i\omega}), v(e^{i\omega}) \rangle \, d\omega,$$

and, by a similar reasoning to part (i), we see that this holds if and only if $G(e^{i\omega})^*G(e^{i\omega}) = I$ a.e. The analogue for the half-plane is now obvious. $\square$

In some sense a linear system with graph $\mathcal{G}(P)$ given by (4.4) has the transfer function NM^{-1}, since an input $u = Mv$ is mapped to an output $y = Nv = NM^{-1}u$. Let us take this discussion a little further, although for simplicity of exposition we shall later restrict to the case $m = p = 1$. Let $\mathcal{F}$ denote the *field of fractions* of the integral domain H^∞ (to be defined on the disc or right half-plane, according to the context). That is, $\mathcal{F}$ consists of all formal quotients f/g with $f, g \in H^\infty$ and $g \not\equiv 0$, where we regard f_1/g_1 to be equivalent to f_2/g_2 whenever $f_1 g_2 = f_2 g_1$. Then, to say that P has $F \in \mathcal{F}$ as its transfer function means that

$$\mathcal{G}(P) = \{(u, Fu) : u, Fu \in H^\infty\}.$$

By a convenient abuse of notation we shall often identify P (an operator) with its transfer function. A matricial factorization $F = NM^{-1}$ over H^∞ is said to be *strongly (right) coprime* if there exist matrix-valued H^∞ functions $\widetilde{X}$ and $\widetilde{Y}$ of the appropriate sizes such that the *Bézout identity* $\widetilde{X}M - \widetilde{Y}N = I$ is satisfied. We then call NM^{-1} a *right coprime factorization* of F (or P). For example, the transfer function $F(s) = 1/s$ arising in Example 4.1.2 has a coprime factorization given by $M(s) = s/(s+1)$ and $N(s) = 1/(s+1)$, since the Bézout identity is satisfied with $\widetilde{X} = 1$ and $\widetilde{Y} = -1$.

The celebrated corona theorem (due to Carleson) implies that two functions f, g in $H^\infty(\mathbb{D})$ are strongly coprime if and only if the *corona condition* condition

$$\inf_{z \in \mathbb{D}} |f(z)| + |g(z)| > 0 \qquad (4.6)$$

holds. We shall not give the proof of this theorem (which is technical, despite simplifications by various authors, notably Hörmander and Wolff), but it can be found, for example, in [2, 39, 92]. The analogous result holds for $H^\infty(\mathbb{C}_+)$.

The following result is elementary, but appealing.

Proposition 4.1.5 *Let a linear system P have a strongly right coprime factorization $F = NM^{-1}$ in terms of matrices over H^∞. Then $\mathcal{G}(P) = \{(Mu, Nu) : u \in H^2(\mathbb{C}^m)\}$.*

Proof: If $v \in \mathcal{D}(P)$, then both $v = MM^{-1}v$ and $NM^{-1}v$ have entries in H^2. Then $u = M^{-1}v = \widetilde{X}MM^{-1}v - \widetilde{Y}NM^{-1}v$ is also defined over H^2 and clearly $(v, Pv) = (Mu, Nu)$.

Conversely, if $u \in H^2(\mathbb{C}^m)$, then (Mu, Nu) is easily seen to lie in the graph of P. $\square$

Theorem 4.1.6 (Smith) *Let a linear system have the transfer function $P \in \mathcal{F}$, and suppose that it is stabilizable by a controller with the transfer function $K \in \mathcal{F}$. Then P has a coprime factorization over H^∞.*

Proof: Recall that if f, g, $h \in H^\infty$, then h is said to be a *greatest common divisor* (GCD) of f and g if it is a common divisor of f and g and a multiple of every other common divisor. We claim first that every two non-constant functions in H^∞ have a GCD.

To see this, let f and g be such functions, and recall from Section 1.3 that they have factorizations $f = f_B f_s f_o$ and $g = g_B g_s g_o$, where f_B is a Blaschke product, f_s is a singular inner function (without zeroes), and f_o is outer (likewise for g). We construct $h = \mathrm{GCD}(f, g)$ as $h = h_B h_s h_o$, where

1. The Blaschke factor h_B has a zero set (counted according to multiplicity) equal to the intersection of the zero sets of f_B and g_B. (Of course, if this is empty, we take $h_B = 1$.)

2. The singular measure μ_h occurring in the expression

$$h_s(re^{it}) = \exp\left(-\frac{1}{2\pi} \int_0^{2\pi} \frac{e^{i\omega} + re^{it}}{e^{i\omega} - re^{it}} \, d\mu_h(\omega)\right) \qquad (4.7)$$

is the maximal measure such that $\mu_h(E) \leq \mu_f(E)$ and $\mu_h(E) \leq \mu_g(E)$ for all Borel sets $E \subseteq \mathbb{T}$ (here μ_f and μ_g are the measures defining f_s and g_s, respectively). The fact that such a μ_h exists can be shown by using the Radon–Nikodym theorem (see the exercises).

3. The outer factor h_o is determined pointwise a.e. by

$$|h_o(e^{i\omega})| = \max\{|f_o(e^{i\omega})|, |g_o(e^{i\omega})|\} = \max\{|f(e^{i\omega})|, |g(e^{i\omega})|\}.$$

Note that this implies that h_o, f_o/h_o and g_o/h_o all lie in H^∞.

It is now clear that $h = h_B h_s h_o$ is a common divisor of f and g, and it is easily verified that any common divisor k of f and g has a factorization $k = k_B k_s k_o$ with k_B a common divisor of f_B and g_B, and similarly for k_s and k_o. It is now straightforward to verify that k_B divides h_B, and similarly for k_s and k_o. Thus k divides h.

Now write $P = N/M$ and $K = Y/X$ with N, M, X and Y in H^∞. By the above, we may suppose that N and M have no non-trivial common factors, and similarly for X and Y. The closed-loop stability implies that the following functions all lie in H^∞:

$$\frac{XM}{XM - YN}, \quad \frac{YM}{XM - YN}, \quad \frac{XN}{XM - YN}, \quad \text{and} \quad \frac{YN}{XM - YN}.$$

Since $XM - YN$ divides XM and YM, it divides their GCD, namely, M, so $M/(XM - YN) \in H^\infty$, and likewise, $N/(XM - YN) \in H^\infty$.

Now $XM - YN$ divides both M and N, so it divides the constant functions, that is, $1/(XM - YN) \in H^\infty$. Thus $\widetilde{X}M - \widetilde{Y}N = I$, where $\widetilde{X} = X/(XM - YN)$ and $\widetilde{Y} = Y/(XM - YN)$ lie in H^∞. □

An extension of Theorem 4.1.6 to the matricial case can be found in [124].

Definition 4.1.7 *A doubly coprime factorization of a matrix-valued function P with entries in the field of fractions of H^∞ is an identity $P = \widetilde{M}^{-1}\widetilde{N} = NM^{-1}$, where*

$$\begin{pmatrix} \widetilde{X} & -\widetilde{Y} \\ -\widetilde{N} & \widetilde{M} \end{pmatrix} \begin{pmatrix} M & Y \\ N & X \end{pmatrix} = \begin{pmatrix} M & Y \\ N & X \end{pmatrix} \begin{pmatrix} \widetilde{X} & -\widetilde{Y} \\ -\widetilde{N} & \widetilde{M} \end{pmatrix} = \begin{pmatrix} I & 0 \\ 0 & I \end{pmatrix}, \quad (4.8)$$

with M, N, $\widetilde{M}$, $\widetilde{N}$, X, Y, $\widetilde{X}$, $\widetilde{Y}$ all being matrix-valued with H^∞ entries and M and N having the same number of columns, with $\widetilde{M}$ and $\widetilde{N}$ having the same number of rows.

In particular we have the Bézout identities $\widetilde{X}M - \widetilde{Y}N = I$ and $\widetilde{M}X - \widetilde{N}Y = I$.

From now on we shall assume that our plants and controllers all have doubly coprime factorizations. We next proceed to derive the Youla parametrization of all stabilizing controllers of P.

Theorem 4.1.8 *Let P have the doubly coprime factorization given in Definition 4.1.7. Then P is stabilizable, and the set of stabilizing controllers K can be parametrized in terms of a right coprime factorization*

$$K = (Y + MQ)(X + NQ)^{-1}$$

and by a left coprime factorization

$$K = (\widetilde{X} + Q\widetilde{N})^{-1}(\widetilde{Y} + Q\widetilde{M}),$$

where in each case Q is an arbitrary matrix of the appropriate size with entries in H^∞.

Proof: It is easy to verify that $K = (Y + MQ)(X + NQ)^{-1}$ is a stabilizing controller for P.

Now let K be a stabilizing controller with a right coprime factorization $K = VU^{-1}$ over H^∞ and write $D = \widetilde{M}U - \widetilde{N}V$.

Now stability implies that $(I - PK)^{-1}$ has entries in H^∞ and

$$(I - PK)^{-1} = (I - \widetilde{M}^{-1}\widetilde{N}VU^{-1})^{-1} = (\widetilde{M}^{-1}(\widetilde{M}U - \widetilde{N}V)U^{-1})^{-1} = UD^{-1}\widetilde{M}.$$

Also, $K(I - PK)^{-1}$ has entries in H^∞, and this is $VD^{-1}\widetilde{M}$. Next,

$$(I - KP)^{-1} = I + K(I - PK)^{-1}P,$$

and so $K(I - PK)^{-1}P$ has entries in H^∞; but this is just $VD^{-1}\widetilde{N}$. Finally,

$$P(I - KP)^{-1} = (I - PK)^{-1}P = UD^{-1}\widetilde{N},$$

and this has entries in H^∞.

Thus all four functions $UD^{-1}\widetilde{M}$, $VD^{-1}\widetilde{M}$, $UD^{-1}\widetilde{N}$ and $VD^{-1}\widetilde{N}$ lie in H^∞. Using the right coprimeness of U and V (say $AU + BV = I$), we conclude that $D^{-1}\widetilde{M} = A(UD^{-1}\widetilde{M}) + B(VD^{-1}\widetilde{M})$ has entries in H^∞, and similarly for $D^{-1}\widetilde{N}$; using the left coprimeness of $\widetilde{M}$ and $\widetilde{N}$, we see that D^{-1} has entries in H^∞.

We define $Q = (\widetilde{X}V - \widetilde{Y}U)D^{-1}$, another matrix over H^∞. Then

$$(Y + MQ)D = Y(\widetilde{M}U - \widetilde{N}V) + M(\widetilde{X}V - \widetilde{Y}U) = V,$$

since $Y\widetilde{M} - M\widetilde{Y} = 0$ and $M\widetilde{X} - Y\widetilde{N} = I$, by (4.8). Similarly,

$$(X + NQ)D = X(\widetilde{M}U - \widetilde{N}V) + N(\widetilde{X}V - \widetilde{Y}U) = U,$$

since $X\widetilde{M} - N\widetilde{Y} = I$ and $-X\widetilde{N} + N\widetilde{X} = 0$, again by (4.8). Thus $K = VU^{-1} = (VD^{-1})(UD^{-1})^{-1} = (Y + MQ)(X + NQ)^{-1}$.

The proof of the left coprime factorization is similar, and we leave it as an exercise. $\square$

There is an attractive geometric way of interpreting feedback stabilization in terms of the graphs of the plant P and its controller K. Recall that the *reversed graph* of an operator $T : \mathcal{H} \to \mathcal{K}$ is defined by $\mathcal{G}'(T) = \{(x, y) \in \mathcal{K} \times \mathcal{H} : x = Ty\}$.

Theorem 4.1.9 *Let P and K be linear systems, as defined above. Then K stabilizes P if and only if*

$$\mathcal{G}(P) \oplus \mathcal{G}'(K) = H^2(\mathbb{C}^{m+p}).$$

Proof: The feedback system $[P, K]$ is stable when the operator on $H^2(\mathbb{C}^{m+p})$ with matrix $\begin{pmatrix} I & -K \\ -P & I \end{pmatrix}$ has a bounded inverse; thus every vector in $H^2(\mathbb{C}^{m+p})$ can be decomposed uniquely in the form

$$\begin{pmatrix} f_1 \\ f_2 \end{pmatrix} = \begin{pmatrix} I & -K \\ -P & I \end{pmatrix} \begin{pmatrix} g_1 \\ g_2 \end{pmatrix} = \begin{pmatrix} g_1 \\ -Pg_1 \end{pmatrix} + \begin{pmatrix} -Kg_2 \\ g_2 \end{pmatrix},$$

with $g_1 \in \mathcal{D}(P)$ and $g_2 \in \mathcal{D}(K)$. The minus signs are not important, as we may change the sign of f_2; thus

$$\begin{pmatrix} f_1 \\ -f_2 \end{pmatrix} = \begin{pmatrix} g_1 \\ Pg_1 \end{pmatrix} + \begin{pmatrix} -Kg_2 \\ -g_2 \end{pmatrix} \in \mathcal{G}(P) + \mathcal{G}'(K).$$

Conversely, if (4.1.9) holds, then the operator $T = \begin{pmatrix} I & -K \\ -P & I \end{pmatrix}$ is invertible, at least algebraically. Its graph $\mathcal{G}(T)$ has four components, which we may write as a column vector,

$$\mathcal{G}(T) = \left\{ \begin{pmatrix} e_1 \\ e_2 \\ e_1 - Ke_2 \\ e_2 - Pe_1 \end{pmatrix} : e_1 \in \mathcal{D}(P), e_2 \in \mathcal{D}(K) \right\}.$$

Since $\mathcal{G}(P)$ and $\mathcal{G}(K)$ are closed, by hypothesis, we conclude that $\mathcal{G}(T)$ is closed, and hence $\mathcal{G}(T^{-1})$ is closed, and T^{-1} is bounded by the closed graph theorem. $\square$

4.2   Robustness

In this section we look at various ways of measuring the distance between two linear systems. In Section 2.3 we introduced the gap metric between two possibly unbounded operators A and B in terms of the gap between their graphs, namely,

$$\delta(A, B) = \delta(\mathcal{G}(A), \mathcal{G}(B)),$$

where, for two closed subspaces $\mathcal{V}$, $\mathcal{W}$ of a Hilbert space, we have

$$\delta(\mathcal{V}, \mathcal{W}) = \|P_{\mathcal{V}} - P_{\mathcal{W}}\|.$$

Our first aim is to show that the gap topology is the appropriate topology to use in the theory of linear systems. For closed operators P that are not necessarily shift-invariant, it still makes sense to talk about coprime factorizations. It is enough to suppose that $\mathcal{D}(P)$ is dense and $\mathcal{G}(P) = \{(Mu, Nu) : u \in \mathcal{H}\}$, where M and N are bounded operators that are strongly right coprime in the sense that $\widetilde{X}M - \widetilde{Y}N = I$ for some operators $\widetilde{X}$ and $\widetilde{Y}$. We continue to write $P = NM^{-1}$.

Proposition 4.2.1 *Suppose that $P = NM^{-1}$ is a right coprime factorization of an operator; then there exists $\epsilon > 0$ such that, if $\|N_1 - N\| < \epsilon$ and $\|M_1 - M\| < \epsilon$, then $P = N_1 M_1^{-1}$ is still a right coprime factorization.*

Proof: We suppose that $\widetilde{X}M - \widetilde{Y}N = I$; then

$$(\widetilde{X}M_1 - \widetilde{Y}N_1 - I) < (\|\widetilde{X}\| + \|\widetilde{Y}\|)\epsilon < 1$$

for sufficiently small ϵ. Then $W = \widetilde{X}M_1 - \widetilde{Y}N_1$ is invertible, and $\widetilde{X}_1 M_1 - \widetilde{Y}_1 N_1 = I$, where $\widetilde{X}_1 = W^{-1}\widetilde{X}$ and $\widetilde{Y}_1 = W^{-1}\widetilde{Y}$.   □

We now want to look at the gap metric distance between P and P_1.

Proposition 4.2.2 *Let $P = NM^{-1}$ and $P_1 = N_1 M_1^{-1}$ be as in Proposition 4.2.1. Then $\delta(P_1, P) \to 0$ as $\epsilon \to 0$. Conversely, for any $\epsilon > 0$ there is an $\eta > 0$ such that any P_1 with $\delta(P_1, P) < \eta$ possesses a coprime factorization $P_1 = N_1 M_1^{-1}$ with $\|N_1 - N\| < \epsilon$ and $\|M_1 - M\| < \epsilon$.*

Proof: Suppose that $(Mu, Nu) \in \mathcal{G}(P)$ with $\|(Mu, Nu)\| = 1$. Then $u = \widetilde{X}Mu - \widetilde{Y}Nu$, and so $\|u\| \leq \max(\|\widetilde{X}\|, \|\widetilde{Y}\|) = C$, say. Now

$$\operatorname{dist}((M_1 u, N_1 u), \mathcal{G}(P)) \leq \|(M_1 u, N_1 u) - (MuNu)\| \leq \sqrt{2}C\epsilon,$$

and thus $\vec{\delta}(\mathcal{G}(P_1), \mathcal{G}(P)) < \sqrt{2}C\epsilon$ (see (2.13)). A similar inequality holds with P and P_1 interchanged, provided that we obtain a bound on $\max(\|\widetilde{X}_1\|, \|\widetilde{Y}_1\|)$. But if $\|W - I\| < \gamma < 1$, then

$$\|W^{-1}\| \leq \sum_{k=0}^{\infty} \|(W - I)^k\| < 1/(1 - \gamma),$$

and thus $\max(\|\widetilde{X}_1\|, \|\widetilde{Y}_1\|) < C/(1-\gamma)$. It now follows from Theorem 2.3.2 that $\delta(P_1, P) \to 0$ as $\epsilon \to 0$.

Conversely, if $\mathcal{V}$ and $\mathcal{W}$ are any two closed subspaces of a Hilbert space with $\delta(\mathcal{V}, \mathcal{W}) < 1$, then the operator $T = P_{\mathcal{V}} P_{\mathcal{W}|\mathcal{V}}$ is an isomorphism from $\mathcal{V}$ onto itself since $\|I - T\| < 1$. Thus $P_{\mathcal{W}|\mathcal{V}}$ is injective and $P_{\mathcal{V}|\mathcal{W}}$ is surjective. Likewise, $P_{\mathcal{W}} P_{\mathcal{V}|\mathcal{W}}$ is an isomorphism of $\mathcal{W}$ onto itself. We conclude that the operator $U = P_{\mathcal{W}|\mathcal{V}}$ provides an isomorphism from $\mathcal{V}$ onto $\mathcal{W}$. Thus, writing $T : \mathcal{G}(P) \to \mathcal{G}(P_1)$ for the isomorphism from $\mathcal{G}(P)$ onto $\mathcal{G}(P_1)$, we see that

$$\mathcal{G}(P_1) = \{T(Mu, Nu) : u \in \mathcal{H}\} = \{(M_1 u, N_1 u) : u \in \mathcal{H}\}, \quad \text{say.}$$

Now M_1 and N_1 are still coprime since

$$(\widetilde{X} \quad -\widetilde{Y}) T^{-1} \begin{pmatrix} M_1 \\ N_1 \end{pmatrix} = I.$$

$\square$

There is a particular form of coprime factorization that will be of use to us, and that is the following.

Definition 4.2.3 *A right coprime factorization* $P = NM^{-1}$ *is said to be norm-alized if* $\begin{pmatrix} M \\ N \end{pmatrix}$ *is an isometry, that is,* $M^*M + N^*N = I$.

In the case when M and N are H^∞ functions, regarded as inducing multiplication operators on H^2, this just means that $\begin{pmatrix} M \\ N \end{pmatrix}$ is inner.

For example, the factorization (in terms of $H^\infty(\mathbb{C}_+)$ functions) $1/s = NM^{-1}$, where $N(s) = 1/(s+1)$ and $M(s) = s/(s+1)$ is normalized, because if $f \in H^2(\mathbb{C}_+)$, then the H^2 norm of the function $s \mapsto (f(s)/(s+1), f(s)s/(s+1))$ is given by

$$\left\| \left(\frac{f(s)}{(s+1)}, \frac{f(s)s}{(s+1)} \right) \right\|^2 = \int_{-\infty}^{\infty} \left(\frac{|f(iy)|^2}{1+y^2} + \frac{|f(iy)|^2 y^2}{1+y^2} \right) dy$$
$$= \|f\|^2.$$

Clearly this holds because for $s = iy$ we have

$$|N(s)|^2 + |M(s)|^2 = 1.$$

Recall that, by Theorem 3.2.5, any closed shift-invariant operator $P : \mathcal{D}(P) \to H^2(\mathbb{C}^p)$ with $\mathcal{D}(T) \subseteq H^2(\mathbb{C}^m)$ satisfies

$$\mathcal{G}(P) = \begin{pmatrix} M \\ N \end{pmatrix} H^2(\mathbb{C}^r) = \Theta H^2(\mathbb{C}^r)$$

for some $r \leq m$, $M \in H^\infty(\mathcal{L}(\mathbb{C}^r, \mathbb{C}^m))$ non-singular, and $N \in H^\infty(\mathcal{L}(\mathbb{C}^r, \mathbb{C}^p))$ such that $\|\Theta u\| = \|u\|$ for all $u \in H^2(\mathbb{C}^r)$. Here $\Theta = \begin{pmatrix} M \\ N \end{pmatrix}$ is a *graph symbol* for P.

The following theorem shows that the calculation of the gap metric can be expressed as an H^∞ optimization problem. For clarity, we suppress the notation $\mathbb{C}^m$, and so on, from now on, although everything is potentially vector-valued.

Theorem 4.2.4 *Let P_1 and P_2 be closed shift-invariant systems with the same number of inputs and outputs and normalized right coprime factorizations $P_j = N_j M_j^{-1}$ and graphs $\mathcal{G}_j$ with graph symbols Θ_j for $j = 1, 2$. Then*

$$\delta(P_1, P_2) = \max \left\{ \inf_{Q \in H^\infty} \|\Theta_1 - \Theta_2 Q\|_\infty, \inf_{Q \in H^\infty} \|\Theta_2 - \Theta_1 Q\|_\infty \right\}. \qquad (4.9)$$

Proof: We know from Theorem 2.3.2 that

$$\delta(P_1, P_2) = \max\{\|(I - P_{\mathcal{G}_2})P_{\mathcal{G}_1}\|, \|(I - P_{\mathcal{G}_1})P_{\mathcal{G}_2}\|\}.$$

Now, by Theorem 3.4.3 (an application of the commutant lifting theorem), we know that

$$\inf_{Q \in H^\infty} \|\Theta_1 - \Theta_2 Q\|_\infty = \|P_\mathcal{X} M_{\Theta_1}\|,$$

where $P_\mathcal{X} = I - P_{\mathcal{G}_2}$ and M_{Θ_1} is the usual multiplication operator. This is easily seen to be the same as the norm of $I - P_{\mathcal{G}_2}$ restricted to $\Theta_1 H^2(\mathbb{C}^r)$, which is the same as $\|(I - P_{\mathcal{G}_2})P_{\mathcal{G}_1}\|$. A similar identity holds when we exchange Θ_1 and Θ_2, and this completes the proof. $\square$

The following alternative metric was introduced by Vidyasagar [134].

Definition 4.2.5 *Let P_1 and P_2 be closed shift-invariant systems with the same number of inputs and outputs and graph symbols Θ_j for $j = 1, 2$. Then the* graph metric $d(P_1, P_2)$ *is defined by*

$$d(P_1, P_2) = \max \left\{ \inf_{Q \in H^\infty, \|Q\|_\infty \leq 1} \|\Theta_1 - \Theta_2 Q\|_\infty, \inf_{Q \in H^\infty, \|Q\|_\infty \leq 1} \|\Theta_2 - \Theta_1 Q\|_\infty \right\}.$$

It is somewhat easier to prove that d is a metric than to prove that δ is, starting from (4.9) as a definition of δ. However, the actual computation of δ in specific examples is a simpler problem, and as a result d is less frequently used these days. In fact, the two metrics are uniformly equivalent.

Theorem 4.2.6 *We have $\delta(P_1, P_2) \leq d(P_1, P_2) \leq 2\delta(P_1, P_2)$ for all P_1 and P_2.*

Proof: The first inequality is obvious, from (4.9), as it is a question of minimizing over larger sets of functions Q.

Suppose now that $Q \in H^\infty$ and $\|\Theta_1 - \Theta_2 Q\|_\infty = \alpha$. Then

$$\|Q\|_\infty \leq \|\Theta_2^* \Theta_1 - Q\|_\infty + \|\Theta_2^* \Theta_1\|_\infty \leq \alpha + 1,$$

since $\Theta_2^* \Theta_2 = I$. Hence

$$\begin{aligned} \|\Theta_1 - \Theta_2 Q/(1+\alpha)\|_\infty &= \|\Theta_1 - \Theta_2 Q + \alpha \Theta_2 Q/(1+\alpha)\|_\infty \\ &\leq \|\Theta_1 - \Theta_2 Q\|_\infty + \|\alpha \Theta_2 Q/(1+\alpha)\|_\infty \leq 2\alpha. \end{aligned}$$

Since $\|Q/(1+\alpha)\|_\infty \leq 1$, we see that $d(P_1, P_2) \leq 2\|\Theta_1 - \Theta_2 Q\|_\infty$, and the result follows by taking the infimum over Q (and using the corresponding inequality with Θ_1 and Θ_2 interchanged). $\qquad\qquad\qquad\qquad\qquad\qquad\square$

Let us now look briefly at *robust* stabilization. The idea here is that, if a system P is stabilized by a controller K, then we should hope that K would also stabilize systems that were close to P in the gap metric. We have seen in Proposition 4.2.2 that if P is expressed by means of a right coprime factorization $P = NM^{-1}$ over H^∞, then the nearby systems in the gap metric all have right coprime factorizations of the form $P_1 = N_1 M_1^{-1}$ with $\|N_1 - N\|_\infty$ and $\|M_1 - M\|_\infty$ small.

We shall keep this discussion at an elementary level; for fuller details and multivariable generalizations, we refer the interested reader to the texts cited in the notes. Accordingly, let us work in the single-input single-output (SISO) case, as follows. At this point it may be helpful for the reader to recall Theorem 4.1.8, which explains the context of the following result.

Theorem 4.2.7 *Let P be the transfer function of a SISO linear system, with a coprime factorization $P = NM^{-1}$ over H^∞. Let K be a stabilizing controller of the form $VU^{-1} = (Y + MQ)(X + NQ)^{-1}$, where X, Y and Q lie in H^∞ and $XM - YN = 1$. Let $b > 0$. Then K stabilizes every perturbed plant $P = N_1 M_1^{-1} = (N + \Delta N)(M + \Delta M)^{-1}$ with $\|(\Delta M, \Delta N)\|_\infty < b$ if and only if*

$$\left\| \begin{pmatrix} Y + MQ \\ X + NQ \end{pmatrix} \right\|_\infty \leq \frac{1}{b}. \tag{4.10}$$

Proof: We have seen already that the stability of $[P_1, K]$ is equivalent to the invertibility of

$$\begin{aligned} D_1 := M_1 U - N_1 V &= (M + \Delta M)(X + NQ) - (N + \Delta N)(Y + MQ) \\ &= 1 + (\Delta M)(X + NQ) - (\Delta N)(Y + MQ). \end{aligned}$$

Clearly, if $\|(\Delta M, \Delta N)\|_\infty < b$ and (4.10) holds, then

$$1 + (\Delta M)(X + NQ) - (\Delta N)(Y + MQ)$$

remains invertible. On the other hand, supposing that (4.10) fails to hold, then we can find a point $s \in \mathbb{C}_+$ for which

$$|(Y + MQ)(s)|^2 + |(X + NQ)(s)|^2 > 1/b^2,$$

and we can make $D_1(s) = 0$ by a suitable scalar choice of $(\Delta M, \Delta N)$ with $|\Delta M|^2 + |\Delta N|^2 < b^2$; this destabilizes $[P_1, K]$. $\square$

This indicates that the gap topology is the appropriate topology in which to consider feedback stabilization. The largest b for which $[P_1, K]$ is stable whenever $\|(\Delta M, \Delta N)\|_\infty < b$ is called the *robustness margin*, and we shall write it b_{opt}.

Corollary 4.2.8 *Suppose that the hypotheses of Theorem 4.2.7 hold and that, in addition, the coprime factorization $P = NM^{-1}$ is normalized. Then the optimal robustness margin b_{opt} is given by*

$$b_{\mathrm{opt}}^{-1} = \inf_{Q \in H^\infty} \left\| \begin{pmatrix} Y \\ X \end{pmatrix} + \begin{pmatrix} M \\ N \end{pmatrix} Q \right\|_\infty = (1 + \|\Gamma_R\|^2)^{1/2},$$

where $\Gamma_R : H^2(\mathbb{C}_+) \to H^2(\mathbb{C}_-)$ is the Hankel operator given by $u \mapsto P_{H^2(\mathbb{C}_-)}(R.u)$, for $u \in H^2(\mathbb{C}_+)$, with $R = M^*Y + N^*X \in L^\infty(i\mathbb{R})$.

Proof: All that remains to be shown is the final identity involving Γ_R. Let

$$Z = \begin{pmatrix} M^* & N^* \\ -N & M \end{pmatrix},$$

so that

$$\left\| \begin{pmatrix} Y + MQ \\ X + NQ \end{pmatrix} \right\|_\infty = \left\| Z \begin{pmatrix} Y + MQ \\ X + NQ \end{pmatrix} \right\|_\infty = \left\| \begin{pmatrix} R + Q \\ I \end{pmatrix} \right\|_\infty.$$

The result now follows from Nehari's theorem, Theorem 3.4.4, in its half-plane version (see the exercises). $\square$

It can be shown (see [41]) that the optimal robustness margin in the gap metric is also equal to b_{opt}, that is, that the following theorem holds.

Theorem 4.2.9 *Let $P = NM^{-1}$ have a normalized coprime factorization, and let K be a stabilizing controller. For a real number $b \in (0, 1]$ the following conditions are equivalent:*

1. *$[P_1, K]$ is stable for all $P_1 = (N + \Delta N)(M + \Delta M)^{-1}$ with $\Delta M, \Delta N \in H^\infty$ and $\left\| \begin{pmatrix} \Delta M \\ \Delta N \end{pmatrix} \right\|_\infty < b$;*

2. *$[P_1, K]$ is stable for all P_1 with $\delta(P, P_1) < b$.*

We shall omit the proof of Theorem 4.2.9 and continue with an example.

Example 4.2.10 Let $P(s) = \dfrac{1}{s-1}$, an unstable system since $P \notin H^\infty(\mathbb{C}_+)$. The function P has coprime factorizations $P = NM^{-1}$ over H^∞, for example,

$$N(s) = \frac{1}{s+1}, \qquad \text{and} \quad M(s) = \frac{s-1}{s+1}.$$

To parametrize all the stabilizing controllers, we do not need to normalize the factorization, and we shall work directly with N and D and the Bézout identity $XM - YN = 1$, where

$$X(s) = 1, \qquad \text{and} \quad Y(s) = -2, \qquad \text{for all } s.$$

Thus the Youla parametrization of all stabilizing controllers gives

$$K(s) = \frac{Y + MQ}{X + NQ}(s) = \frac{-2 + \frac{s-1}{s+1}Q(s)}{1 + \frac{1}{s+1}Q(s)}, \qquad Q \in H^\infty(\mathbb{C}_+).$$

There are virtues in simplicity, especially when we are doing worked examples, so let us take $Q = 0$, giving $K(s) \equiv -2$ ("constant negative feedback", in the control jargon), so that the closed-loop operator $H(P, K)$ given in Proposition 4.1.3 is given by

$$H(K, P)(s) = \begin{pmatrix} (I - KP)^{-1} & K(I - PK)^{-1} \\ P(I - KP)^{-1} & (I - PK)^{-1} \end{pmatrix}(s) = \begin{pmatrix} \frac{s-1}{s+1} & -2\frac{s-1}{s+1} \\ \frac{1}{s+1} & \frac{s-1}{s+1} \end{pmatrix},$$

all of whose entries are in $H^\infty(\mathbb{C}_+)$. The robustness margin b is given by

$$b^{-1} = \left\| \begin{pmatrix} Y + MQ \\ X + NQ \end{pmatrix} \right\|_\infty = \left\| \begin{pmatrix} -2 \\ 1 \end{pmatrix} \right\|_\infty = \sqrt{5},$$

and to see how this applies, consider the perturbed system $N_1 M_1^{-1}$ where

$$(N_1, M_1)(s) = (N, M)(s) - \left(\frac{2}{5}, \frac{1}{5} \right),$$

where the perturbation has $\|(\Delta N, \Delta M)\| = 1/\sqrt{5}$. We find that

$$P_1(s) = \frac{\frac{1}{s+1} - \frac{2}{5}}{\frac{s-1}{s+1} - \frac{1}{5}}$$

and $(I - KP_1)(s) \equiv 0$, so the feedback system $[P_1, K]$ is unstable.

To find the optimal robustness margin, we repeat some of the calculation using normalized coprime factorizations. At this point our calculations will be simplified by the observation that if F is a real rational function, then $F^*(s) = F(-s)$ for $s \in i\mathbb{R}$, since $\overline{F(iy)} = F(-iy)$ for $y \in \mathbb{R}$. Thus, in our original coprime factorization, we have

$$(N^*N + M^*M)(s) = \frac{2 - s^2}{1 - s^2},$$

from which it is not hard to see that a *normalized* coprime factorization is given by $P = NM^{-1}$, where

$$N(s) = \frac{1}{s + \sqrt{2}}, \qquad \text{and} \quad M(s) = \frac{s - 1}{s + \sqrt{2}}.$$

From now on we work with this choice of N and M; the Bézout identity is satisfied if we take $X(s) \equiv 1$ and $Y(s) \equiv -1 - \sqrt{2}$. To calculate the robustness margin b_{opt} we calculate $R = M^*Y + N^*X$, or

$$R(s) = \frac{(\sqrt{2} + 1)s + (2 + \sqrt{2})}{\sqrt{2} - s}.$$

Using the fact that constant functions lie in H^∞ and that for $a > 0$ the distance of $1/(s - a)$ to $H^\infty(\mathbb{C}_+)$ is $1/(2a)$ (see the exercises), it is not hard to verify that $\|\Gamma_R\| = \text{dist}(R, H^\infty(\mathbb{C}_+)) = 1 + \sqrt{2}$. We conclude that $b_{\text{opt}} = (4 + 2\sqrt{2})^{-1/2}$.

We have so far ignored two basic practical questions:

1. How can we construct Bézout identities? and

2. How can we construct normalized coprime factorizations?

The first of these is rather difficult in general. We present a construction for delay systems in Section 6.4, but even this is fairly complicated; for the moment we shall discuss the case of *rational* functions. Recall that P is *proper*, if the degree of its denominator is greater than or equal to the degree of its numerator.

Proposition 4.2.11 *Let P be a proper rational function. Then it has a coprime factorization $P = N/M$ in terms of rational $H^\infty(\mathbb{C}_+)$ functions.*

Proof: Let $P = p/q$, where p and q are polynomials with no common factor such that $\deg q = n \geq 1$ and $\deg p \leq n$. Choose any polynomial r of degree n such that r has no zeroes in the closed right half-plane, for example, $r(s) = (s + 1)^n$. Our coprime factors will be $N = p/r$ and $M = q/r$, but to show that they are coprime we need to establish a Bézout identity.

By means of the Euclidean algorithm it is possible to find polynomials u and v such that $up + vq = 1$. Indeed we may suppose without loss of generality that

u and v both have degree at most $n - 1$, since, if not, we may write $u = wq + t$ with w, t polynomials and $\deg t < n$; then $tp + (v + wp)q = (wq + t)p + vq = 1$ and $\deg(v + wp) + \deg q = \deg t + \deg p$, so that t and $v + wp$ both have degree at most $n - 1$.

Now $r^2 = xq + y$ for polynomials x and y with $\deg x = n$ and $\deg y < n$. Now y too can be written as $ap + bq$ for polynomials a, b with $\deg a < n$ and $\deg b < n$, for $y = (uy)p + (vy)q$, and, repeating the argument of the previous paragraph, we have $uy = cq + d$, where $\deg c \leq (2n - 2) - n = n - 2$ and $\deg d < n$; thus $y = dp + (vy + cp)q$ and $\deg vy + cp \leq n - 1$, since $\deg dp \leq 2n - 1$.

We arrive at the identity $r^2 = (x + vy + cp)q + dp$, and thus

$$\frac{d}{r}\left(\frac{p}{r}\right) + \frac{x + vy + cp}{r}\left(\frac{q}{r}\right) = 1,$$

and all four fractions are in $H^\infty(\mathbb{C}_+)$. $\qquad\qquad\square$

Example 4.2.12 Let $P(s) = s/(s-1)^2$, and take $N(s) = s/(s+1)^2$ and $M(s) = (s-1)^2/(s+1)^2$. The Euclidean algorithm (or inspection) gives us

$$-(s-2)s + 1(s-1)^2 = 1,$$

and division gives

$$(s+1)^4 = (s^2 + 6s + 17)(s^2 - 2s + 1) + 32s - 16,$$

from which we arrive directly at

$$(s+1)^4 = (s^2 + 38s + 1)(s-1)^2 - 16(2s^2 - 5s + 2)s,$$

indicating that $XM - YN = 1$, where

$$X(s) = \frac{s^2 + 38s + 1}{(s+1)^2} \qquad \text{and} \quad Y(s) = \frac{16(2s^2 - 5s + 2)}{(s+1)^2}.$$

In general, such calculations are best left to a computer.

While in a practical frame of mind, let us consider the second problem: how to construct a normalized coprime factorization. Again, we shall restrict discussion to the scalar rational case. The key is in the following result, which is a version of the classical Fejér–Riesz theorem.

Theorem 4.2.13 *Let N and M be rational functions in $H^\infty(\mathbb{C}_+)$ such that*

$$\inf_{y \in i\mathbb{R}} |N(iy)|^2 + |M(iy)|^2 > 0.$$

Then there is a rational function $F \in H^\infty(\mathbb{C}_+)$ with $1/F \in H^\infty(\mathbb{C}_+)$ such that $|F(iy)|^2 = |N(iy)|^2 + |M(iy)|^2$ for $y \in \mathbb{R}$. Moreover, $(N/F), (M/F)$ is a normalized coprime factorization.

Proof: Let $R(s) = N(s)\overline{N(-\bar{s})} + M(s)\overline{M(-\bar{s})}$. This is a rational function of s, and it is positive and bounded both above and below on the imaginary axis. We may assume without loss of generality that R is in its lowest terms.

Now the zeroes and poles of R occur in pairs symmetric about the y-axis, since R is identical with the rational function $\widetilde{R}(s) = \overline{R(-\bar{s})}$, because they coincide on the imaginary axis. Thus if $w \in \mathbb{C}$ and $R(w) = 0$, then $R(-\overline{w}) = 0$ too (and similarly for poles). We may therefore write $R(s) = F(s)G(s)$, where

$$F(s) = c \prod_{j=1}^{m} \frac{s - z_j}{s - p_j}, \quad \text{and} \quad G(s) = c \prod_{j=1}^{m} \frac{s + \overline{z_j}}{s + \overline{p_j}},$$

for some $c > 0$, with $z_1, \ldots, z_m$ and $p_1, \ldots, p_m$, respectively, the zeroes and poles of F in the left half-plane (it is easy to see that their numbers are equal). Thus $R(s) = F(s)\overline{F(-\bar{s})}$, and on the imaginary axis

$$|N(iy)|^2 + |M(iy)|^2 = R(iy) = |F(iy)|^2.$$

It is now clear that N/F and M/F give a normalized coprime factorization.   □

Example 4.2.14 Let us take $N(s) = s/(s+1)^2$ and $M(s) = (s-1)^2/(s+1)^2$, as in Example 4.2.12. We need only work with the numerators and calculate

$$\begin{aligned}
(s)(-s) + (s-1)^2(-s-1)^2 &= s^4 - 3s^2 + 1 \\
&= (s^2 + \sqrt{5}s + 1)(s^2 - \sqrt{5}s + 1),
\end{aligned}$$

and the roots of $s^2 + \sqrt{5}s + 1$ both lie in the left half-plane. We therefore have the following normalized coprime factors:

$$\frac{s}{s^2 + \sqrt{5}s + 1} \quad \text{and} \quad \frac{(s-1)^2}{s^2 + \sqrt{5}s + 1}.$$

The above process is known as a *spectral factorization* of R. More generally, the inner–outer factorization of an H^∞ function allows one to construct (or at least give a formula for) a spectral factorization of an arbitrary positive invertible function on $L^\infty(i\mathbb{R})$, as follows.

Theorem 4.2.15 *Let $R \in L^\infty(i\mathbb{R})$ be such that $R > 0$ a.e. and $1/R \in L^\infty(i\mathbb{R})$. Then there is a function $F \in H^\infty(\mathbb{C}_+)$ with the property that $|F(iy)|^2 = R(iy)$ a.e. Thus, if N and M are functions in $H^\infty(\mathbb{C}_+)$ with*

$$\operatorname*{ess\,inf}_{y \in i\mathbb{R}} |N(iy)|^2 + |M(iy)|^2 > 0,$$

then there exists a function $F \in H^\infty(\mathbb{C}_+)$ with $1/F \in H^\infty(\mathbb{C}_+)$ such that

$$|F(iy)|^2 = |N(iy)|^2 + |M(iy)|^2$$

a.e. for $y \in \mathbb{R}$. Moreover, $(N/F), (M/F)$ is a normalized coprime factorization over $H_\infty(\mathbb{C}_+)$.

Proof: It follows from Equation (1.7) that the function F, defined by

$$F(s) = \exp\left(\frac{1}{2\pi}\int_{-\infty}^{\infty}\frac{ys+i}{y+is}\log|R(iy)|\frac{dy}{1+y^2}\right),$$

is an outer function in $H^\infty(\mathbb{C}_+)$ such that $\log|F(iy)| = \frac{1}{2}\log|R(iy)|$ almost everywhere. The remaining deductions are straightforward. $\square$

This is an existence theorem, and the formula above is not normally used in explicit calculations. It may be remarked at this point that spectral factorization, and the construction of coprime factorizations in general, presents various technical difficulties. For example, Treil [131] gave an example to show that functions with continuous boundary values need not have continuous spectral factors. This implies that the process of spectral factorization is discontinuous in the uniform norm, and some care has to be exercised in working with approximate coprime factorizations. We refer to [18] and [62] for further details.

4.3 The chordal metric

We now turn our attention to a topic with its roots in classical complex analysis, which allows one to apply function-theoretic methods to robust control.

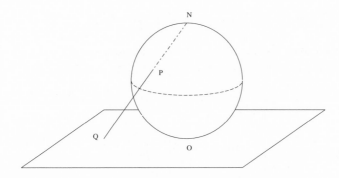

Figure 4.2. The Riemann sphere and stereographic projection

Recall that the process of stereographic projection provides a way of identifying the extended complex plane $\mathbb{C} \cup \{\infty\}$ with the sphere of unit diameter S

in $\mathbb{R}^3$ (the *Riemann sphere*). In Figure 4.2, the point $z = 0$ corresponds to the south pole O of the sphere of unit diameter, and the point $z = \infty$ corresponds to its north pole N. We then identify points Q in the plane with points P on the sphere whenever N, P and Q lie on a straight line. It is not hard to see that this correspondence is continuous, at least between $\mathbb{C}$ and $S \setminus \{N\}$, and indeed it induces a natural metric on $\mathbb{C} \cup \{\infty\}$, as follows.

We define the chordal distance between two points in $\mathbb{C} \cup \{\infty\}$ by measuring the length of the chord between the corresponding points on the Riemann sphere.

For two complex numbers w_1 and w_2, the chordal distance between them on the Riemann sphere is

$$\kappa(w_1, w_2) = \frac{|w_1 - w_2|}{\sqrt{(1 + |w_1|^2)(1 + |w_2|^2)}},$$

with $\kappa(w, \infty) = 1/\sqrt{1 + |w|^2}$.

Definition 4.3.1 *For two meromorphic functions P_1, P_2 in the open right half-plane, the* chordal distance κ *between P_1 and P_2 is defined by*

$$\kappa(P_1, P_2) = \sup\{\kappa(P_1(s), P_2(s)) \; : \; \operatorname{Re} s > 0\}.$$

The chordal metric is sometimes referred to as the *pointwise gap metric*, and this is because

$$\kappa(P_1, P_2) = \sup\{\delta(P_1(s), P_2(s)) : \operatorname{Re} s > 0\},$$

where we interpret $P_1(s)$ and $P_2(s)$ as linear operators from $\mathbb{C}$ to $\mathbb{C}$. (This leads to an obvious multivariable generalization.) Note that, with this convention,

$$\mathcal{G}(P(s)) = \begin{cases} \{(a, P(s)a) : a \in \mathbb{C}\} & \text{if } P(s) \text{ is finite,} \\ \{(0, b) : b \in \mathbb{C}\} & \text{if } P(s) = \infty, \end{cases}$$

and the computation of the gap between two one-dimensional subspaces of $\mathbb{C}^2$ is an easy exercise (indeed, it was Exercise 14 of Chapter 2).

The following straightforward inequality will be useful to us.

Proposition 4.3.2 *Let w_1 and w_2 be points in $\mathbb{C}$. Then*

$$\kappa(w_1, w_2) \leq \min\left(|w_1 - w_2|, \left|\frac{1}{w_1} - \frac{1}{w_2}\right|\right). \tag{4.11}$$

Proof: This is elementary, given that $\kappa(w_1, w_2) = \kappa(1/w_1, 1/w_2)$ for all w_1 and w_2 in $\mathbb{C}$. $\square$

In some ways, this metric is a more intuitive one to consider than the graph and gap metrics: one can see at once that two functions P and P' are close if at each point either $|P(s) - P'(s)|$ is small or $|1/P(s) - 1/P'(s)|$ is small – this latter case handles poles of P and P'. It is also useful to have a lower bound for κ, which acts as a partial converse to Proposition 4.3.2.

Lemma 4.3.3 *For any two complex numbers w_1 and w_2, and for any a with $0 < a < 1$,*

$$\kappa(w_1, w_2) \geq \frac{1}{1 + a^2} \min\left\{ a^2 |w_1 - w_2|, a^2 \left| \frac{1}{w_1} - \frac{1}{w_2} \right|, 1 - a^2 \right\}.$$

Proof: Consider the following three cases, which are collectively exhaustive:
  (a) $|w_1| \leq 1/a$ and $|w_2| \leq 1/a$;
  (b) $|w_1| \geq a$ and $|w_2| \geq a$;
  (c) either $|w_1| \leq a$ and $|w_2| \geq 1/a$, or vice versa.
In case (a), the formula for κ shows that $\kappa(w_1, w_2) \geq |w_1 - w_2|/(1 + 1/a^2)$; in case (b), the same applies with w_1 and w_2 replaced by $1/w_1$ and $1/w_2$; and in case (c), $\kappa(w_1, w_2) \geq \kappa(a, 1/a) = (1 - a^2)/(1 + a^2)$. $\square$

El-Sakkary [28] gave some early robustness results for the chordal metric, of which the following is typical: if $1/(1 + P)$ is stable and

$$\kappa(P(s), P'(s)) < \frac{1}{3(1 + |[1 + P(s)]|^{-1}|^2)^{1/2}},$$

then $1/(1 + P')$ is also stable (and there is a bound on $\left| \frac{1}{1+P(s)} - \frac{1}{1+P'(s)} \right|$.) This corresponds to stabilizing P by a constant controller K.

Using Lemma 4.3.3, we can now prove some new robustness results in the chordal metric. In order to keep the calculations at an elementary level, we shall work with a SISO plant P and a fixed controller $K \in H^\infty$. Note that the stability requirement is therefore simply that $P/(1 + KP)$ lie in H^∞, since this implies successively that $PK/(1 + KP)$, $1/(1 + KP)$ and $K/(1 + KP)$ are also in H^∞. Such systems are called *strongly stabilizable*, but in fact this notion coincides with stabilizability (see [130] for this rather deep result, which can be regarded as a generalization of the corona theorem).

Theorem 4.3.4 *Suppose that P_0 and P_1 are SISO transfer functions and K is a stable controller such that the closed-loop transfer function $G_0 = P_0/(1 + KP_0)$ lies in H_∞. Write $k = \|K\|_\infty$ and $g = \|G_0\|_\infty$. If*

$$\kappa(P_0, P_1) \leq (1/3) \min\{1, g^{-1}, k^{-1}(1 + kg)^{-1}\},$$

then P_1 is also stabilized by K, that is, $G_1 = P_1/(1 + KP_1) \in H_\infty$.

Proof: Suppose that s is a point such that $G_1(s) = \infty$ but $|G_0(s)| \leq g = \|G_0\|_\infty < \infty$. We may estimate the chordal distance of $P_0(s)$ from $P_1(s)$ by using Lemma 4.3.3. Let a be any number with $0 < a < 1$. Then

(a) $|P_0(s) - P_1(s)| = \left| \dfrac{1}{K(s)(1 - K(s)G_0(s))} \right| \geq \dfrac{1}{k(1 + kg)}$;   and

(b) $\left| \dfrac{1}{P_0(s)} - \dfrac{1}{P_1(s)} \right| = \left| \dfrac{1}{G_0(s)} \right| \geq \dfrac{1}{g}$.

Thus

$$\kappa(P_0, P_1) \geq \frac{1}{1 + a^2} \min \left\{ \frac{a^2}{k(1 + kg)}, \frac{a^2}{g}, 1 - a^2 \right\}.$$

An easy limiting argument shows that the same is true if G_1 is merely unbounded, rather than having a pole in the right half-plane.

The result now follows on taking $a = 1/\sqrt{2}$. $\square$

Given $k = \|K\|$ and $g = \|G_0\|$, it is in general possible to choose a in order to obtain tighter estimates, but we do not do this here.

We can improve upon the above result and show that if P_0 is close to P_1, then G_0 is close to G_1 in H_∞ norm, as follows.

Theorem 4.3.5 *Under the hypotheses of Theorem 4.3.4, let $\epsilon > 0$ be given. Then, if*

$$\kappa(P_0, P_1) < (1/3) \min \left\{ 1, \frac{\epsilon}{(1 + kg)(1 + k(g + \epsilon))}, \frac{\epsilon}{g(g + \epsilon)} \right\},$$

then $\|G_0 - G_1\|_\infty < \epsilon$.

Proof: Again we use Lemma 4.3.3 and the estimates

$$|P_0(s) - P_1(s)| = \frac{|G_0(s) - G_1(s)|}{|1 - K(s)G_0(s)| \, |1 - K(s)G_1(s)|}$$

and

$$\left| \frac{1}{P_0(s)} - \frac{1}{P_1(s)} \right| = \frac{|G_0(s) - G_1(s)|}{|G_0(s)| \, |G_1(s)|}$$

to bound κ from below, under the hypothesis that $|G_0(s) - G_1(s)| = \epsilon$. $\square$

Systems that are close in the gap topology are also close in the chordal metric, as the following result shows.

Lemma 4.3.6 *Suppose that $P_0(s)$ and $P_1(s)$ have coprime factorizations over H_∞, namely, $P_0(s) = N_0(s)/M_0(s)$ and $P_1(s) = N_1(s)/M_1(s)$. Then*

$$\kappa(P_0, P_1) \leq A \max(\|N_0 - N_1\|, \|M_0 - M_1\|),$$

where A is a constant that can be taken to depend only on P_0, not on P_1.

Proof: Since N_0 and M_0 are coprime, they satisfy a Bézout identity

$$X(s)M_0(s) - Y(s)N_0(s) = 1$$

with $X, Y \in H^\infty$. Let $x = \|X\|_\infty$ and $y = \|Y\|_\infty$. Then certainly for each s we have that $x|M_0(s)| + y|N_0(s)| \geq 1$, and hence if $|M_0(s)| \leq 1/2x$, then $|N_0(s)| \geq 1/2y$. We can thus estimate $\kappa(N_0/M_0, N_1/M_1)$: when $|M_0(s)| \geq 1/2x$,

$$\kappa\left(\frac{N_0}{M_0}, \frac{N_1}{M_1}\right) \leq \left|\frac{N_0}{M_0} - \frac{N_1}{M_1}\right| \leq \frac{|N_0||M_1 - M_0| + |M_0||N_0 - N_1|}{|M_0||M_1|},$$

which will be at most a constant times $\max(\|N_0 - N_1\|_\infty, \|M_0 - M_1\|_\infty)$ provided that $\|M_0 - M_1\|_\infty < 1/4x$; alternatively, when $|M_0(s)| < 1/2x$ and $|N_0(s)| \geq 1/2y$,

$$\kappa\left(\frac{N_0}{M_0}, \frac{N_1}{M_1}\right) \leq \left|\frac{M_0}{N_0} - \frac{M_1}{N_1}\right| \leq \frac{|M_0||N_1 - N_0| + |N_0||M_0 - M_1|}{|N_0||N_1|},$$

using Proposition 4.3.2, and this is bounded by a constant times

$$\max(\|N_0 - N_1\|_\infty, \|M_0 - M_1\|_\infty)$$

provided that $\|N_0 - N_1\| < 1/4x$. $\qquad\square$

To avoid technical complications we give the next result in its simplest form only.

Theorem 4.3.7 *The chordal metric gives the gap topology when restricted to the set of proper rational functions.*

Proof: By Proposition 4.2.2 and Lemma 4.3.6, a sequence (P_n) of proper rational transfer functions that converges to another one, P, in the gap metric will also converge in the chordal metric.

Conversely, if $P_n \to P$ in the chordal metric, then, letting $p_1, \ldots, p_m$ be the poles of P in $\{\operatorname{Re} s \geq 0\}$, the following is a coprime factorization of P:

$$N(s) = G(s)\prod_{j=1}^{m}\frac{s - p_j}{s + 1}, \qquad \text{and} \quad M(s) = \prod_{j=1}^{m}\frac{s - p_j}{s + 1},$$

and any function P_n sufficiently close to P in the chordal metric will have precisely m poles $p_{n,1}, \ldots, p_{n,m}$ with $p_{n,j}$ close to p_j for each j, by Rouché's theorem, since $P_n^{-1} \to P^{-1}$ uniformly on a neighbourhood of each p_j and $P_n \to P$ uniformly on the complement of the union of these neighbourhoods. We now have $P_n = N_n/M_n$, where

$$N_n(s) = G_n(s)\prod_{j=1}^{m}\frac{s - p_{n,j}}{s + 1}, \qquad \text{and} \quad M_n(s) = \prod_{j=1}^{m}\frac{s - p_{n,j}}{s + 1}.$$

It is clear that $\|M_n - M\|_\infty \to 0$, and it follows easily that $\|N_n - N\|_\infty \to 0$ as well. Hence we have convergence in the gap metric, by Proposition 4.2.2.   □

There is a multivariable version of the chordal metric, the ν-gap metric, described in [138]. By imposing an additional winding number condition it is possible to reduce the calculation of the chordal metric to an optimization over the imaginary axis, rather than the entire half-plane, which is clearly a computational advantage.

Notes

The literature on the automatic continuity of shift-invariant operators and casual operators on various spaces is extensive, and we refer the reader to [21, 71, 74, 85, 123].

Theorem 4.1.6 is taken from [124], where it answers a question posed in [137]. The result holds in the matrix-valued case, too, and one obtains strongly coprime left and right factorizations. See also [135] for more on the link between stabilization and coprime factorization.

Carleson's original proof of the corona theorem is in [15].

The Youla parametrization may be found in many places, for example [20, 37, 135, 151]. The original sources are [147, 148].

The link between feedback stability and the structure of the graphs of the plant and controller was discovered almost simultaneously by several authors. See [35, 42, 94, 153].

For our discussion of the gap metric and robustness margins, we have drawn on [41, 86, 94, 121, 122, 136].

Theorems 4.2.4 and 4.2.6 are due to Georgiou [40]. Theorem 4.2.7 is due to Vidyasagar and Kimura [136], whereas Corollary 4.2.8 is from Glover and Mc-Farlane [86].

More on the chordal metric can be found in the book of Hayman [54].

Some papers developing properties of the chordal metric (pointwise gap metric) are [27, 28, 98, 99, 112, 150]. Theorem 4.3.4 and related results are taken from [99].

It has been asserted that the chordal metric actually coincides with the gap metric when restricted to systems with no unstable zeroes or poles (cf. [27]), although in fact this is not always the case [41].

Exercises

1. Prove the discrete-time version of Theorem 4.1.1, namely, that any linear shift-invariant operator on $\ell^2(\mathbb{Z}_+)$ is bounded.

2. Verify that the operator of multiplication by $1/s$ (defined on a subdomain of $H^2(\mathbb{C}_+)$) is unitarily equivalent to the operator T in Example 4.1.2.

3. Verify that Equation (4.2) has solutions given by Equation (4.3) when all the necessary inverse operators exist.

4. Prove the polarization identity (4.5).

5. Let μ_f and μ_g be positive Borel measures, as in the proof of Theorem 4.1.6. By using the Radon–Nikodym theorem to write $d\mu_f = F d\mu$ and $d\mu_g = G d\mu$, where $\mu = \mu_f + \mu_g$, show that there is a maximal Borel measure μ_h such that $\mu_h(E) \leq \min\{\mu_f(E), \mu_g(E)\}$ for all Borel sets E.

6. Derive the formula for the left coprime Youla parametrization, as given in Theorem 4.1.8.

7. Prove that the graph metric is indeed a metric.

8. Deduce Nehari's theorem for $H^\infty(\mathbb{C}_+)$, either directly from the commutant lifting theorem or from the analogous version for the disc (Theorem 3.4.4).

9. Show that if $a > 0$ and $f_a(s) = 1/(s - a)$, then $\mathrm{dist}(f_a, H^\infty(\mathbb{C}_+)) = \dfrac{1}{2a}$, and that a closest point is the constant function $-1/2a$.

10. Calculate an optimally robust controller for Example 4.2.10 by solving the appropriate Nehari problem, using Exercise 9.

11. Repeat the calculations of Example 4.2.10 for the unstable system

$$P(s) = \frac{1}{s}.$$

12. Calculate coprime factorizations, with corresponding Bézout identities, for the systems $P_1(s) = \dfrac{1}{s^2 + 1}$ and $P_2(s) = \dfrac{1}{s^2 - 4}$.

13. Calculate normalized coprime factorizations for the systems P_1 and P_2 given in Exercise 12.

14. The *model matching problem* for a plant P is to minimize the quantity $\|P(I - KP)^{-1} - R\|_\infty$, where $R \in H^\infty$ is given, over all choices of stabilizing controllers K. Use the Youla parametrization to reformulate this as an H^∞ minimization problem in terms of the free parameter Q.

15. Repeat the calculation of Exercise 14 for the *sensitivity minimization problem*, where now it is required to minimize $\|(I - KP)^{-1}\|_\infty$.

16. Show that any function of the form

$$\tau(z) = e^{i\theta} \frac{az + b}{-\bar{b}z + \bar{a}},$$

with $a,\, b \in \mathbb{C}$ and $\theta \in \mathbb{R}$, is a conformal isometry of the Riemann sphere, that is, it satisfies $\kappa(\tau(w_1), \tau(w_2)) = \kappa(w_1, w_2)$ for all w_1, w_2. Show also that any point w_1 can be mapped to any other point w_2 by using these maps.

Chapter 5

Spaces of persistent signals

So far we have worked almost entirely with signal spaces of the form $\ell^2(\mathbb{Z}_+)$ or $L^2(0, \infty)$ and their full-axis analogues; in physical terms, these are spaces of *finite-energy signals*, which die away in some sense at infinity. In this chapter we shall work with what may loosely be described as *finite-power signals* or, still more loosely, as *persistent signals*.

Persistent signals include classes of signals with some regularity properties, such as periodic and almost-periodic signals, as well as much more general spaces of signals in which the notion of "power" is less clearly defined. In particular, we are are able to discuss concepts such as the idea of a white noise signal in a rigorous and largely non-stochastic framework. Persistent signals in general can be taken as the inputs and outputs of linear systems (the term *filter* is commonly used here), as we shall see.

5.1 Almost-periodic functions

Almost-periodic functions defined on the real line form a class of functions that has been much studied since the 1920s. Our aim in this section is to derive their fundamental properties and to bring out their similarities with the theory of periodic functions.

We begin by recalling the basic properties of 2π-periodic functions on $\mathbb{R}$. These are totally standard results and may be found in [69, 149, 155] and many other places. If $F : \mathbb{R} \to \mathbb{C}$ is 2π-periodic (i.e., $F(t) = F(t + 2\pi)$ for all $t \in \mathbb{R}$) and Lebesgue integrable over finite intervals, then it may also be regarded as a function in $L^1(\mathbb{T})$ by means of the correspondence

$$f(e^{it}) = F(t), \qquad (t \in \mathbb{R}).$$

It corresponds to a complex sequence of Fourier coefficients given by

$$\hat{f}(n) = \frac{1}{2\pi} \int_0^{2\pi} f(e^{it})e^{-int}\, dt, \qquad (n \in \mathbb{Z}).$$

If $f \in L^2(\mathbb{T})$, then we have

$$f(e^{it}) = \sum_{n=-\infty}^{\infty} \hat{f}(n)e^{int}, \tag{5.1}$$

where convergence is interpreted in the $L^2(\mathbb{T})$ norm. We also have *Parseval's identity:* if $f \in L^2(\mathbb{T})$, then

$$\frac{1}{2\pi} \int_0^{2\pi} |f(e^{it})|^2\, dt = \sum_{n=-\infty}^{\infty} |\hat{f}(n)|^2.$$

If $f \in C(\mathbb{T})$, then we cannot guarantee pointwise convergence of the Fourier series (5.1), let alone uniform convergence, but we can recover f by means of its *Fejér sums*, namely,

$$f_n(e^{it}) = \sum_{k=-n}^{n} \left(1 - \frac{|k|}{n+1}\right) \hat{f}(k)e^{ikt}, \qquad (n = 1, 2, \ldots), \tag{5.2}$$

which converge uniformly to f whenever f is continuous. If $\sum_{n=-\infty}^{\infty} |\hat{f}(n)| < \infty$, then the Fourier series is said to be *absolutely convergent*, and it converges uniformly to a continuous function. Returning to the real line, we see that every continuous 2π-periodic function is the uniform limit of a sequence of trigonometric polynomials. (Sometimes one plays this game with the functions $\sin nt$ and $\cos nt$ rather than e^{int}, but there is no essential difference.)

Let us now mimic the above results, working with the full set of functions $\{e_\lambda : \lambda \in \mathbb{R}\}$ given by
$$e_\lambda(t) = e^{i\lambda t}, \qquad (t \in \mathbb{R}),$$
whereas previously we restricted λ to lying in $\mathbb{Z}$. The definition we give now is not totally standard (although it has been used elsewhere, e.g., [19]), but, as we shall see later, it is equivalent to Bohr's original definition.

Definition 5.1.1 *The class $AP(\mathbb{R})$ of (uniformly) almost-periodic functions is the closed linear span in $L^\infty(\mathbb{R})$ of the set of functions $(e_\lambda)_{\lambda \in \mathbb{R}}$.*

Since the uniform limit of continuous functions is continuous, we see that $AP(\mathbb{R}) \subseteq C_b(\mathbb{R})$, the space of continuous bounded functions on $\mathbb{R}$. However, an almost-periodic function is not, in general, periodic (see the exercises).

For example, Figure 5.1 shows a plot of the function $2\sin t + \cos \sqrt{2}t$.

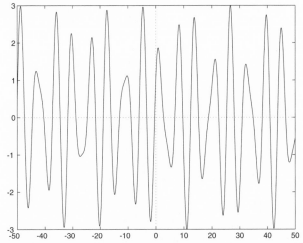

Figure 5.1. The almost-periodic function $2\sin t + \cos\sqrt{2}t$

Recall that the linear space $L^2_{\mathrm{loc}}(\mathbb{R})$ consists of all Lebesgue-integrable functions $f : \mathbb{R} \to \mathbb{C}$ such that

$$\int_A^B |f(t)|^2\, dt < \infty \qquad \text{for all } A, B \in \mathbb{R}.$$

We introduce the following inner-product-like notation for a pair of functions $f, g \in L^2_{\mathrm{loc}}(\mathbb{R})$ and $T > 0$:

$$[f, g]_T = \frac{1}{2T} \int_{-T}^{T} f(t)\overline{g(t)}\, dt \tag{5.3}$$

and, when the limit exists,

$$[f, g] = \lim_{T \to \infty} [f, g]_T. \tag{5.4}$$

Then we have the following orthogonality relation:

$$[e_\lambda, e_\mu] = \begin{cases} 1 & \text{if } \lambda = \mu, \\ 0 & \text{if } \lambda \neq \mu, \end{cases} \tag{5.5}$$

as the reader is invited to verify.

Remark 5.1.2 It is possible to define a Hilbert space by taking the completion of the set of trigonometric polynomials under the norm $[f, f]^{1/2}$. This yields a space that is sometimes called $AP_2(\mathbb{R})$, which has an uncountable orthonormal basis $(e_\lambda)_{\lambda \in \mathbb{R}}$. However, although $AP(\mathbb{R})$ embeds continuously into this space, $AP_2(\mathbb{R})$ is not a space of functions defined on $\mathbb{R}$. We leave the reader to consider what interpretation, if any, can be given to the formal series $\sum_{k=1}^{\infty} \frac{1}{k} \exp(it/k)$, which represents an object in $AP_2(\mathbb{R})$ but where the sum diverges for every $t \in \mathbb{R}$.

To give an intrinsic definition of the class $AP(\mathbb{R})$, we shall make use once more of the right shift R_λ defined for any $f : \mathbb{R} \to \mathbb{C}$ by $(R_\lambda f)(x) = f(x - \lambda)$.

Definition 5.1.3 *Let $\epsilon > 0$ and a continuous function $f : \mathbb{R} \to \mathbb{C}$ be given. A number $\lambda \in \mathbb{R}$ is called an ϵ-translation number of f if*

$$\|R_\lambda f - f\|_\infty \le \epsilon,$$

that is, if $|f(x - \lambda) - f(x)| \le \epsilon$ for all $x \in \mathbb{R}$. A set $\mathcal{S} \subseteq \mathbb{R}$ is said to be relatively dense *if there exists an $L > 0$ such that every interval of length L contains an element of $\mathcal{S}$.*

We now study the class of functions satisfying the following properties, which will turn out to be equivalent to belonging in $AP(\mathbb{R})$.

Definition 5.1.4 *We say that a function $f : \mathbb{R} \to \mathbb{C}$ is a* Bohr function *if f is continuous and for every $\epsilon > 0$ the set of ϵ-translation numbers of f is relatively dense. The function $f : \mathbb{R} \to \mathbb{C}$ satisfies* Bochner's condition *if it is continuous and bounded and the set of translates $\{R_\lambda f : \lambda \in \mathbb{R}\}$ is relatively compact (i.e., totally bounded) in $C_b(\mathbb{R})$.*

For example, if f is a τ-periodic function for some $\tau > 0$, then $n\tau$ is an ϵ-translation number for any $\epsilon > 0$ and $n \in \mathbb{Z}$, and so f is a Bohr function. We are working towards showing that the two conditions above are equivalent, and that the class of functions satisfying them coincides with the class $AP(\mathbb{R})$ defined earlier.

Theorem 5.1.5 *The class of Bohr functions is the same as the class of functions satisfying Bochner's condition.*

Proof: We show first that the Bohr functions are bounded. For given such a function f, let L be such that every interval of length L contains a 1-translation number of f. Let $M = \max\{|f(t)| : t \in [0, L]\}$, and let $x \in \mathbb{R}$ be given. Then there is a 1-translation number λ for f in the interval $[\lambda - L, \lambda]$. Now $|f(x)| \le |f(x - \lambda)| + 1 \le M + 1$, as required.

Moreover, every Bohr function is uniformly continuous. For given such an f and $\epsilon > 0$, let $L > 0$ be such that every interval of length L contains an $\frac{\epsilon}{3}$-translation number of f. Since f is uniformly continuous on $[0, L + 1]$, we may find δ with $0 < \delta < 1$ such that $|f(x_1) - f(x_2)| < \epsilon/3$ whenever $x_1, x_2 \in [0, L+1]$ and $|x_1 - x_2| < \delta$. Given any $y_1, y_2 \in \mathbb{R}$ with $|y_1 - y_2| < \delta$, we may find an $\frac{\epsilon}{3}$-translation number $\tau \in \mathbb{R}$ such that the points $x_1 = y_1 - \tau$ and $x_2 = y_2 - \tau$ lie in $[0, L + 1]$. Now

$$|f(y_1) - f(y_2)| \le |f(y_1) - f(y_1 - \tau)| + |f(y_1 - \tau) - f(y_2 - \tau)| + |f(y_2 - \tau) - f(y_2)| < \epsilon.$$

It follows that we have $\|R_{z_k}f - R_z f\|_\infty \to 0$ whenever (z_k) is a real sequence with $z_k \to z$.

Now suppose, if possible, that f is a Bohr function that is not a Bochner function. Then for some $\epsilon > 0$ we have a sequence (λ_k) such that $\|R_{\lambda_j}f - R_{\lambda_k}f\|_\infty > \epsilon$ for all $j \neq k$. However, there is a number L such that every interval of length L contains an $\frac{\epsilon}{4}$-translation number of f. Write $\lambda_k = \tau_k + \delta_k$, where τ_k is an $\frac{\epsilon}{4}$-translation number and $0 \leq \delta_k \leq L$. Thus $\|R_{\lambda_k}f - R_{\delta_k}f\| \leq \epsilon/4$ for each k. By passing to a subsequence and relabelling, we may suppose that the sequence (δ_k) converges, with limit δ, say, and by uniform continuity we may suppose further that $\|R_{\delta_k}f - R_\delta f\|_\infty < \epsilon/4$ for all k. Thus, if $j \neq k$, we have

$$\begin{aligned} \|R_{\lambda_j}f - R_{\lambda_k}f\|_\infty &\leq \|R_{\lambda_j}f - R_{\delta_j}f\|_\infty + \|R_{\delta_j}f - R_\delta f\|_\infty \\ &\quad + \|R_\delta f - R_{\delta_k}f\|_\infty + \|R_{\delta_k}f - R_{\lambda_k}f\|_\infty < \epsilon, \end{aligned}$$

which is a contradiction and shows that "Bohr implies Bochner".

Conversely, if f is not a Bohr function, there exists an $\epsilon > 0$ for which the set $\mathcal{S}_\epsilon$ of ϵ-translation numbers of f is not relatively dense. Take $c_1 = 1$ and let (a_2, b_2) be an interval of length greater than 2 containing no element of $\mathcal{S}_\epsilon$. Let $c_2 = (a_2 + b_2)/2$. Inductively, define (a_n, b_n) as intervals of length greater than $2(|c_1| + |c_2| + \ldots + |c_{n-1}|)$ containing no element of $\mathcal{S}_\epsilon$ and let $c_n = (a_n + b_n)/2$. If $1 \leq k < n$, then $c_n - c_k \notin \mathcal{S}_\epsilon$. Now

$$\|R_{c_n}f - R_{c_k}f\|_\infty = \|R_{c_n - c_k}f - f\| > \epsilon,$$

and so f does not satisfy the Bochner condition. $\qquad\square$

Using the Bochner condition allows us to see easily that the $AP(\mathbb{R})$ functions lie in the class of Bohr functions.

Corollary 5.1.6 *The Bohr functions form a closed linear subspace of $C_b(\mathbb{R})$, and hence every function in $AP(\mathbb{R})$ is a Bohr function.*

Proof: If f_1 and f_2 are Bohr functions, c_1 and c_2 are complex constants, and (λ_k) is a real sequence, then, by passing to a subsequence and relabelling, we may suppose without loss of generality that $(R_{\lambda_k}f_1)$ and $(R_{\lambda_k}f_2)$ are convergent sequences in $C_b(\mathbb{R})$. It now follows easily that $(R_{\lambda_k}(c_1 f_1 + c_2 f_2))$ is a convergent sequence, and this establishes that $c_1 f_1 + c_2 f_2$ is a Bohr function.

Moreover, the class of Bohr functions is closed, since, if λ is an $\frac{\epsilon}{3}$-translation number for f and $\|f - g\|_\infty < \epsilon/3$, then

$$\|R_\lambda g - g\|_\infty \leq \|R_\lambda g - R_\lambda f\|_\infty + \|R_\lambda f - f\|_\infty + \|f - g\|_\infty < \epsilon,$$

and so λ is an ϵ-translation number for g. Thus, if g is in the closure of the set of Bohr functions, we can find a relatively dense set of ϵ-translation numbers for g by taking a relatively dense set of $\epsilon/3$ translation numbers for any Bohr function f with $\|f - g\|_\infty < \epsilon/3$. Thus g is also a Bohr function.

We have seen already that every function e_λ is a Bohr function, since it is periodic. Hence $AP(\mathbb{R})$, the closed linear span of the e_λ, consists entirely of Bohr functions. $\qquad\square$

Remark 5.1.7 It is easily verified, in addition, that the Bohr functions form a closed subalgebra of $C_b(\mathbb{R})$. All that remains is to show that the pointwise product $f_1 \cdot f_2$ of two Bohr functions is a Bohr function. The proof proceeds as in the first part of Corollary 5.1.6), using the observation that if $(R_{\lambda_k} f_1)$ and $(R_{\lambda_k} f_2)$ are norm-convergent, then so is $(R_{\lambda_k}(f_1.f_2))$.

We now wish to prove a rather deep result, namely, that every Bohr function is a limit of trigonometric polynomials, that is, finite linear combinations of the functions e_λ. To do this, we need to identify which frequencies are present in an almost-periodic oscillation, and here the indefinite inner-product formulae $[f, g]_T$ and $[f, g]$ defined in (5.3) and (5.4) are required once more.

Proposition 5.1.8 *Let f be a Bohr function. Then*

$$[f, 1] = \lim_{T \to \infty} \frac{1}{2T} \int_{-T}^{T} f(t)\, dt$$

exists. Hence

$$[f, g] = \lim_{T \to \infty} \frac{1}{2T} \int_{-T}^{T} f(t)\overline{g(t)}\, dt$$

is well defined for all Bohr functions f and g.

Proof: Suppose that $T > 0$ and let $M = \|f\|_\infty$. Given $\epsilon > 0$, let L be such that every interval of length L contains an ϵ-translation number of f. Write

$$\frac{1}{2nT} \int_{-nT}^{nT} f(t)\, dt = \frac{1}{2nT} \sum_{k=-n}^{n-1} \int_{kT}^{(k+1)T} f(t)\, dt, \qquad (5.6)$$

and suppose that $T > \max\{L, ML/\epsilon\}$. Then, if τ is an ϵ-translation number of f in $[kT, kT + L]$, we have

$$\int_{kT}^{(k+1)T} f(t)\, dt = \int_{0}^{T} f(t)\, dt + \int_{0}^{T} (f(t + \tau) - f(t))\, dt$$
$$+ \int_{kT}^{\tau} f(t)\, dt - \int_{(k+1)T}^{T+\tau} f(t)\, dt,$$

and we may bound the second, third and fourth terms on the right-hand side, using the "size of function" times "length of interval" rule, by ϵT, ML and ML, respectively. Hence, adding up the $2n$ terms in (5.6) we obtain

$$|[f,1]_{nT} - [f,1]_T| \leq \epsilon + \frac{ML}{T} + \frac{ML}{T} \leq 3\epsilon.$$

In particular, $[f,1]_{nT}$ remains bounded. For $U > 0$ sufficiently large, choose n such that $nT \leq U < (n+1)T$. Then

$$
\begin{aligned}
|[f,1]_U - [f,1]_{nT}| &\leq \left|[f,1]_U - \frac{nT}{U}[f,1]_{nT}\right| + \left|\frac{nT}{U} - 1\right| |[f,1]_{nT}| \\
&\leq \frac{1}{2U}2MT + \frac{1}{n}|[f,1]_{nT}| \\
&\leq \frac{M}{n} + \frac{1}{n}|[f,1]_{nT}| < \epsilon
\end{aligned}
$$

if U is sufficiently large. Thus

$$|[f,1]_U - [f,1]_T| < 4\epsilon$$

if U is sufficiently large. This gives $|[f,1]_U - [f,1]_V| < 8\epsilon$ when U and V are sufficiently large, implying the existence of the limit $[f,1]$.

Note also that $[f,g] = [f\bar{g},1]$ is defined for all Bohr functions f and g, since the product $f \cdot \bar{g}$ is also a Bohr function, by Remark 5.1.7. $\qquad\square$

The mapping from $\mathbb{R}$ to $\mathbb{C}$ defined by $\lambda \mapsto [f,e_\lambda]$ is sometimes called the *Bohr transform*. The numbers $[f,e_\lambda]$ are called the *Fourier coefficients* of f.

We shall find it useful to introduce an auxiliary function, the *correlation* or *covariance function* of two Bohr functions f_1 and f_2. This is defined for $x \in \mathbb{R}$ by

$$\phi_{f_1,f_2}(x) = [R_{-x}f_1, f_2] = \lim_{T \to \infty} \frac{1}{2T} \int_{-T}^{T} f_1(x+t)\overline{f_2(t)}\, dt. \qquad (5.7)$$

We shall see more of this idea in Section 5.3.

Proposition 5.1.9 *Let f_1 and f_2 be Bohr functions; then so is the covariance function $\phi = \phi_{f_1,f_2}$. Moreover, $[R_{-x}f_1, f_2]_T \to \phi_{f_1,f_2}(x)$ uniformly in x as $T \to \infty$. Also, $[R_{-x}f_1, f_2] = [f_1, R_x f_2]$.*

Proof: Since

$$\phi(x-\lambda) - \phi(x) = \lim_{T \to \infty} \frac{1}{2T} \int_{-T}^{T} (f_1(x+t-\lambda) - f_1(x+t))\overline{f_2(t)}\, dt,$$

we see easily that

$$\|R_\lambda \phi - \phi\|_\infty \leq \|R_\lambda f_1 - f_1\|_\infty \|f_2\|_\infty,$$

from which we conclude that ϕ is also a Bohr function, since a δ-translation number for f_1 is an ϵ-translation number for ϕ as soon as $\delta \|f_2\|_\infty \leq \epsilon$.

For any fixed x, the convergence of $[R_{-x}f_1, f_2]_T$ to $\phi(x)$ is clear from Proposition 5.1.8. Now, given $\epsilon > 0$, we may use the Bochner property of f_1 to find $x_1, \ldots, x_n \in \mathbb{R}$ such that for each $x \in \mathbb{R}$ there is a k with $\|R_{-x}f_1 - R_{-x_k}f_1\|_\infty < \epsilon$. Thus

$$|[R_{-x}f_1, f_2]_T - [R_{-x_k}f_1, f_2]_T| < \epsilon \|f_2\|_\infty$$

for all $T > 0$. Moreover, there is a number T_0 such that

$$|[R_{-x_k}f_1, f_2]_T - [R_{-x_k}f_1, f_2]| < \epsilon$$

for all $T \geq T_0$, for all the finite collection $k = 1, \ldots, n$. Now the triangle inequality implies that if $T \geq T_0$, then

$$|[R_{-x}f_1, f_2]_T - [R_{-x}f_1, f_2]| < \epsilon(1 + 2\|f_2\|_\infty)$$

for all x.

Note that

$$
\begin{aligned}
[R_{-x}f_1, f_2]_T &= \frac{1}{2T}\int_{-T}^{T} f_1(x+t)\overline{f_2(t)}\, dt \\
&= \frac{1}{2T}\int_{-T+x}^{T+x} f_1(s)\overline{f_2(s-x)}\, ds \\
&= [f_1, R_x f_2]_T + \delta(T),
\end{aligned}
$$

where $|\delta(T)| \leq |x|\|f_1\|_\infty \|f_2\|_\infty / T$, and so $[R_{-x}f_1, f_2] = [f_1, R_x f_2]$. $\square$

Since the convergence of $[R_{-x}f, 1]_T$ to $[f, 1]$ is uniform in x, the quantity $[f, 1]$ can be estimated by integrating over any sufficiently long interval $[-T+x, T+x]$, or, what is the same, using $[R_{-x}f, 1]_T$ for large T.

Using these tools, we are now ready to perform some harmonic analysis on the class of Bohr functions, which will lead to a proof that it coincides with $AP(\mathbb{R})$.

Note that the mapping $(f, g) \mapsto [f, g]$ satisfies all the axioms for an inner product on the class of Bohr functions except possibly the positive definiteness condition (in fact, it satisfies this too, but we shall defer this to later and merely use semi-definiteness for now). This enables us to develop an inner-product space theory of almost-periodic functions in a simple manner. Recall that, in any inner-product space, if $u_1, u_2, \ldots, u_n$ is an orthonormal sequence and x is any vector, then setting

$$u = \sum_{k=1}^{n} \langle x, u_k \rangle u_k,$$

we have that $x - u$ is orthogonal to every u_k, and hence it is orthogonal to u itself. Pythagoras's theorem now gives

$$\|x\|^2 = \|x - u\|^2 + \|u\|^2 \geq \|u\|^2 = \sum_{k=1}^{n} |\langle x, u_k \rangle|^2, \qquad (5.8)$$

which is *Bessel's inequality*. Moreover, if v is an arbitrary linear combination of $u_1, \ldots, u_n$, then $(x - u) \perp (u - v)$, and so

$$\|x - v\|^2 = \|(x - u) + (u - v)\|^2 = \|x - u\|^2 + \|u - v\|^2 \geq \|x - u\|^2,$$

which is the "best approximation" property of u.

Theorem 5.1.10 *Let f be a Bohr function. Then Bessel's inequality holds, in the following form*

$$[f, f] \geq \sum_{k=1}^{n} |[f, e_{\lambda_k}]|^2$$

for all distinct $\lambda_1, \ldots, \lambda_n$ in $\mathbb{R}$. Hence $[f, e_\lambda] \neq 0$ for at most a countable set of $\lambda \in \mathbb{R}$.

Proof: We obtain Bessel's inequality directly from (5.8), writing f for x and e_{λ_k} for u_k. This implies that, for any $N \geq 1$, we have $|[f, e_\lambda]| > 1/N$ for at most finitely many λ. Hence the total number of non-zero Fourier coefficients $[f, e_\lambda]$ is at most countable. $\qquad \square$

We can now prove that $(f, g) \mapsto [f, g]$ is a genuine inner product on the Bohr functions, that is, it is positive definite.

Theorem 5.1.11 *Let f be a Bohr function that is not identically zero. Then $[f, f] > 0$.*

Proof: If f is not identically zero, then there is an $\epsilon > 0$ and an $a \in \mathbb{R}$ such that $|f(a)| > \epsilon$. By continuity, we may find $\delta > 0$ such that $|f| \geq \epsilon/2$ on the interval $(a - \delta, a + \delta)$.

Let L be such that every interval of length L contains an $\frac{\epsilon}{4}$-translation number of f. Then for $n \geq 1$ we have

$$\frac{1}{2nL} \int_{-nL}^{nL} |f(t)|^2 \, dt \geq \frac{\delta \epsilon^2}{16L},$$

since each interval $[kL - a, (k+1)L - a]$ contains a $\frac{\epsilon}{4}$-translation number τ_k of f, implying that $|f| \geq \epsilon/4$ on $(a + \tau_k - \delta, a + \tau_k + \delta) \cap [kL, (k+1)L]$; this has length at least δ since $a + \tau_k \in [kL, (k+1)L]$. Hence $[f, f] \geq \frac{\delta \epsilon^2}{16L}$, as required. $\square$

Remark 5.1.12 Suppose that (f_k) is a sequence of Bohr functions for which $[f_k - f, f_k - f] \to 0$ for some Bohr function f. If an additional property holds, namely, that for each $\epsilon > 0$ every ϵ-translation number of f is an ϵ-translation number of all the functions f_k, then it follows by the same argument as above that (f_k) tends to f *uniformly*. For if $|(f - f_k)(a_k)| > \epsilon$, then we still obtain $|f - f_k| \geq \epsilon/2$ on some interval $(a_k - \delta, a_k + \delta)$, where δ is independent of k, since sufficiently small numbers are $\epsilon/4$-translation numbers for f, and hence for f_k. Now we can find many intervals on which $|f - f_k| \geq \epsilon/4$ and estimate an integral once more.

At last we are ready to prove the fundamental uniqueness theorem, stating that the Fourier coefficients determine the function uniquely, that is, if $[f, e_\lambda] = 0$ for all λ, then f is identically zero. To do this we need a preliminary lemma.

Lemma 5.1.13 *Suppose that f is a Bohr function such that $[f, e_\lambda] = 0$ for all $\lambda \in \mathbb{R}$. Then $[f, e_\lambda]_T \to 0$ as $T \to \infty$ uniformly in λ.*

Proof: We shall suppose the contrary and derive a contradiction. So suppose that $|[f, e_{\lambda_n}]_{T_n}| \geq \epsilon > 0$ for sequences $(\lambda_n) \subset \mathbb{R}$ and T_n tending to ∞. Observe that, for every $\lambda \neq 0$,

$$\begin{aligned}
[f, e_\lambda]_T &= \frac{1}{2T} \int_{-T}^{T} f(t) e^{-i\lambda t} \, dt \\
&= \frac{1}{2T} \int_{-T + \pi/\lambda}^{T + \pi/\lambda} f\left(t - \frac{\pi}{\lambda}\right) e^{-i\lambda(t - \pi/\lambda)} \, dt,
\end{aligned}$$

which implies that

$$[f, e_\lambda]_T = \frac{1}{4T} \int_{-T}^{T} \left(f(t) - f\left(t - \frac{\pi}{\lambda}\right) \right) e^{-i\lambda t} \, dt + \delta(\lambda),$$

where the first term tends to zero uniformly in T as $|\lambda| \to \infty$, by the uniform continuity of f, and

$$\delta(\lambda) = \frac{-1}{4T} \left(\int_{-T + \pi/\lambda}^{-T} + \int_{T}^{T + \pi/\lambda} \right) f\left(t - \frac{\pi}{\lambda}\right) e^{-i\lambda t} \, dt = O(T^{-1} \lambda^{-1}).$$

It follows that the given sequence (λ_n) must remain bounded and has a convergent subsequence. By relabelling we may suppose without loss of generality that $\lambda_n \to \lambda$. Write $\lambda_n = \lambda + \delta_n$, where $\delta_n \to 0$.

By Proposition 5.1.9, $[R_{-s}f, e_\lambda]_U \to [R_{-s}f, e_\lambda] = [f, R_s e_\lambda] = 0$ uniformly in s as $U \to \infty$, so we can find a number $U_0 > 0$ such that $|[R_{-s}f, e_\lambda]_U| < \epsilon/2$ for all $U \geq U_0$ and for all $s \in \mathbb{R}$. Now, given $T_n > U_0$, we may write $T_n = NU$ for some U with $U_0 < U < 2U_0$ and $N \in \mathbb{N}$, both depending on n. Therefore

$$[f, e_{\lambda_n}]_{T_n} = \frac{1}{N} \sum_{j=0}^{N-1} e^{-i\lambda_n((2j+1-N)U)} \frac{1}{2U} \int_{-U}^{U} f(t + (2j+1-N)U)e^{-i\lambda t}e^{-i\delta_n t}\, dt.$$

But $e^{-i\delta_n t} \to 1$ as $n \to \infty$ uniformly for $t \in [-2U_0, 2U_0]$, so that $|[f, e_{\lambda_n}]_{T_n}| < \epsilon$ for sufficiently large n. This is a contradiction, and the result follows. $\square$

Theorem 5.1.14 (Uniqueness theorem) *Let f be a Bohr function such that $[f, e_\lambda] = 0$ for all $\lambda \in \mathbb{R}$. Then f is identically zero.*

Proof: We begin by defining for each $T > 0$ an auxiliary function f_T that equals f on the interval $(-T, T)$ and is $2T$-periodic. Thus f_T has a Fourier series

$$f_T(t) \sim \sum_{k=-\infty}^{\infty} a_k e^{i\pi kt/T},$$

and Parseval's identity gives us

$$\frac{1}{2T} \int_{-T}^{T} |f(t)|^2\, dt = \sum_{k=-\infty}^{\infty} |a_k|^2.$$

The proof now proceeds by working with the quantity $\sum_{k=-\infty}^{\infty} |a_k|^4$, which depends on T; we note that, given $\epsilon > 0$, we have for sufficiently large T that $|a_k| = |[f, e_{\pi k/T}]_T| < \epsilon$ for all k, by Lemma 5.1.13. Thus

$$\sum_{k=-\infty}^{\infty} |a_k|^4 < \epsilon^2 \sum_{k=-\infty}^{\infty} |a_k|^2 \leq \epsilon^2 \|f\|_\infty^2. \tag{5.9}$$

We now construct a new $2T$-periodic function g_T (an *autocorrelation* function) defined by

$$g_T(t) = \frac{1}{2T} \int_{-T}^{T} f_T(t + s)\overline{f_T(s)}\, ds. \tag{5.10}$$

We leave the reader to verify that the Fourier coefficients of g_T are equal to $|a_k|^2$. This can be done by a simple change of order of integration or by an approximation argument based on the relation for finite sums:

$$\frac{1}{2T} \int_{-T}^{T} \left(\sum_{k=-N}^{N} a_k e^{i\pi k(t+s)/T} \right) \left(\sum_{l=-N}^{N} \overline{a_l} e^{-i\pi l s/T} \right) ds = \sum_{k=-N}^{N} |a_k|^2 e^{i\pi k t/T}.$$

To obtain a function with coefficients $|a_k|^4$, it is clearly enough to repeat the construction and define

$$h_T(t) = \frac{1}{2T} \int_{-T}^{T} g_T(t+s)\overline{g_T(s)}\, ds. \tag{5.11}$$

Now $h_T(0) = \sum_{k=-\infty}^{\infty} |a_k|^4$, because the Fourier series of h_T converges absolutely and hence pointwise. This tends to zero as $T \to \infty$, by (5.9), and so $[g_T, g_T]_T \to 0$ as $T \to \infty$ by (5.11). Now take $T_n \to \infty$ such that T_n is a $\frac{1}{n}$-translation number of f and note that, for $0 \le t \le T_n$, we have

$$
\begin{aligned}
g_{T_n}(t) &= \frac{1}{2T_n} \int_{-T_n}^{T_n-t} f(t+s)\overline{f(s)}\, ds + \frac{1}{2T_n} \int_{T_n-t}^{T_n} f(t+s-T_n)\overline{f(s)}\, ds \\
&= \frac{1}{2T_n} \int_{-T_n}^{T_n} f(t+s)\overline{f(s)}\, ds + \delta_n,
\end{aligned}
$$

where $|\delta_n| \le \|f\|_\infty/n$. Clearly the same estimate holds for $-T_n \le t \le 0$.

Recall from Proposition 5.1.9 that the function $g : \mathbb{R} \to \mathbb{C}$ defined by

$$g(t) = [R_{-t}f, f] = \lim_{U \to \infty} \frac{1}{2U} \int_{-U}^{U} f(t+s)\overline{f(s)}\, ds \tag{5.12}$$

is also a Bohr function and that the convergence of the right-hand side of (5.12) to $g(t)$ as $U \to \infty$ is uniform in t.

We see therefore that

$$\eta_n = \sup\{|g_{T_n}(t) - g(t)| : |t| \le T_n\} \to 0 \qquad \text{as } n \to \infty.$$

Moreover,

$$
\begin{aligned}
|[g_{T_n}, g_{T_n}]_{T_n} - [g, g]_{T_n}| &\le \frac{1}{2T_n} \int_{-T_n}^{T_n} (|g_{T_n}(t)| + |g(t)|)\, (|g_{T_n}(t) - g(t)|)\, dt \\
&\le 2\|f\|_\infty^2 \eta_n,
\end{aligned}
$$

which tends to zero as n tends to infinity. Thus $[g, g] = 0$, and so g is identically zero by Theorem 5.1.11. But $g(0) = [f, f]$, and we finally conclude that f is identically zero. $\qquad\square$

Thus for Bohr functions the Fourier coefficients $[f, e_\lambda]$ determine the function f uniquely. We have already seen from Bessel's inequality (Theorem 5.1.10) that $[f, e_\lambda] \neq 0$ for at most a countable set of $\lambda \in \mathbb{R}$.

Theorem 5.1.15 (Parseval's identity) *Let f be a Bohr function. Then one has* $[f, f] = \sum_{\lambda \in \mathbb{R}} |[f, e_\lambda]|^2$.

Proof: We work once more with the almost-periodic covariance function $g = \phi_{f,f}$ and note that

$$[g, e_\lambda] = \lim_{T \to \infty} \lim_{X \to \infty} \frac{1}{2X} \int_{-X}^{X} e^{-i\lambda x} [R_{-x} f, f]_T \, dx,$$

since $[R_{-x} f, f]_T \to g(x)$ uniformly on $\mathbb{R}$, by Proposition 5.1.9. Using Fubini's theorem, this gives

$$
\begin{aligned}
[g, e_\lambda] &= \lim_{T \to \infty} \lim_{X \to \infty} \frac{1}{2X} \int_{-X}^{X} \frac{1}{2T} \int_{-T}^{T} e^{-i\lambda(x+t)} e^{i\lambda t} f(x+t) \overline{f(t)} \, dt \, dx \\
&= \lim_{T \to \infty} \frac{1}{2T} \int_{-T}^{T} [f, e_\lambda] e^{i\lambda t} \overline{f(t)} \, dt = |[f, e_\lambda]|^2.
\end{aligned}
$$

Now $\sum_{\lambda \in \mathbb{R}} |[f, e_\lambda]|^2 < \infty$, by Theorem 5.1.10, and so the series $\sum_{\lambda \in \mathbb{R}} |[f, e_\lambda]|^2 e_\lambda$ converges uniformly to a Bohr function h whose Fourier coefficients satisfy $[h, e_\lambda] = |[f, e_\lambda]|^2$ for all λ (because of the uniform convergence), and hence $g = h$ by the uniqueness theorem, Theorem 5.1.14. Evaluating at $x = 0$, we see that $\sum_{\lambda \in \mathbb{R}} |[f, e_\lambda]|^2 = g(0) = [f, f]$, as required. $\square$

It remains to show that every Bohr function is in $AP(\mathbb{R})$, the closed linear span of the functions e_λ. To do this, we take an arbitrary Bohr function f and consider the set $\Lambda = \{\lambda \in \mathbb{R} : [f, e_\lambda] \neq 0\}$. If this set is finite, then there is no problem, since we form the trigonometric polynomial $h = \sum_{\lambda \in \Lambda} [f, e_\lambda] e_\lambda$. It is clear that h has the same Fourier coefficients as f, and so, by the uniqueness theorem, $f = h$. We may therefore suppose without loss of generality that Λ is countably infinite, say

$$\Lambda = \{\lambda_1, \lambda_2, \ldots\}.$$

The first step in our approximation procedure is to reduce Λ to a maximal subset $B = \{\beta_1, \beta_2, \ldots\}$ (possibly finite) that is linearly independent over $\mathbb{Q}$. This can be done recursively, by successively deleting λ_k if it is a linear combination with rational coefficients of λ_j for $j < k$. Alternatively, some form of Zorn's lemma can be used. We shall assume without loss of generality that B is countably infinite (if not, we extend it to a countably infinite independent set by adding in new members).

For a fixed positive integer n, let E_n be the finite set consisting of all numbers of the form

$$\lambda = \sum_{k=1}^{n} \frac{m_k}{n!} \beta_k,$$

where $m_k \in \mathbb{Z}$ and $|m_k| \leq n.n!$ for each k. It is not hard to see that $\Lambda \subseteq E = \bigcup_{n=1}^{\infty} E_n$, simply because any rational coefficient p/q can be written as $m/n!$ with $|m| \leq n.n!$ if n is sufficiently large.

Definition 5.1.16 *Given a countable set $B = \{\beta_1, \beta_2, \ldots\}$, linearly independent over $\mathbb{Q}$, and a positive integer n, the* Fejér–Bochner kernel K_n' *corresponding to B is given by*

$$K_n'(t) = \prod_{k=1}^{n} K_{n.n!-1}(\beta_k t/n!),$$

where K denotes the "standard" Fejér kernel, defined for $p \in \mathbb{Z}_+$ and $\omega \in \mathbb{R}$ by the formula

$$K_p(\omega) = \sum_{m=-p-1}^{p+1} \left(1 - \frac{|m|}{p+1}\right) e^{im\omega}.$$

We may write $K_n'(t) = \sum_{\lambda \in E_n} k_n(\lambda) e^{i\lambda t}$, in which case we see the following:

1. $K_n'(t) \geq 0$ for all $t \in \mathbb{R}$, since K_n' is a product of "standard" Fejér kernels;

2. $k_n(\lambda) = k_n(-\lambda)$ for each λ, and hence $K_n'(t) = K_n'(-t)$ for all $t \in \mathbb{R}$; also, $k_n(0) = 1$;

3. for each $\lambda \in E$, we have $0 \leq k_n(\lambda) \leq 1$ for all n, and $k_n(\lambda) \to 1$ as $n \to \infty$.

Part 3 holds because, if $\lambda = \sum_{k=1}^{r} m_k \beta_k/r! \in E_r$, then

$$k_n(\lambda) = \prod_{k=1}^{r} \left(1 - \frac{|m_k|}{n.n!}\right)$$

if $n \geq r$, and this tends to 1 as $n \to \infty$.

Using this kernel, we are now able to prove the approximation theorem we wanted.

Theorem 5.1.17 *Given a Bohr function f, let $(f_n)_{n=1}^{\infty}$ denote the sequence of trigonometric polynomials defined by*

$$f_n(x) = [R_{-x}f, K_n'] = [f, R_x K_n'] = \lim_{T \to \infty} \frac{1}{2T} \int_{-T}^{T} f(t) K_n'(t-x) \, dt, \qquad (x \in \mathbb{R}).$$

Then $\|f - f_n\|_\infty \to 0$ as $n \to \infty$ and thus f is almost-periodic.

Proof: We observe first that for $x \in \mathbb{R}$ we have

$$f_n(x) = \sum_{\lambda \in E_n} [f, k_n(\lambda) R_x e_\lambda] = \sum_{\lambda \in E_n} k_n(\lambda)[f, e_\lambda] e_\lambda(x).$$

Using Parseval's identity (Theorem 5.1.15), it is clear by the dominated convergence theorem that $[f - f_n, f - f_n] \to 0$, since $[f - f_n, e_\lambda] \to 0$ for all λ and $|[f - f_n, e_\lambda]| \leq |[f, e_\lambda]|$ for each λ.

We observe that any ϵ-translation number τ pertaining to f is also an ϵ-translation number for f_n, since

$$|f_n(x - \tau) - f_n(x)| = [R_\tau f - f, R_x K_n'] \leq \epsilon \lim_{T \to \infty} \frac{1}{2T} \int_{-T}^{T} K_n'(t - x)\, dt = \epsilon,$$

using the positivity of K_n' and the fact that $k_n(0) = 1$.

The proof is now completed by using Remark 5.1.12.   □

We have now completed the circle of ideas that identifies the Bohr functions with the almost-periodic functions, that is, the uniform limits of trigonometric polynomials. Let us now consider these from the point of view of linear systems and, in particular, convolution operators.

Theorem 5.1.18 *Suppose that $g \in L^1(0, \infty)$. Then the convolution operator T_g given by*

$$(T_g u)(t) = \int_0^\infty g(\tau) u(t - \tau)\, d\tau$$

defines a bounded operator on $AP(\mathbb{R})$, with norm given by $\|T_g\| = \|g\|_{L^1(0,\infty)}$. Moreover, T_g extends by continuity to a bounded operator on the Hilbert space $AP_2(\mathbb{R})$ with norm equal to $\|Lg\|_{H^\infty(\mathbb{C}_+)}$, where L denotes the Laplace transform.

Proof: It is clear that T_g is bounded, even on $C_b(\mathbb{R})$, since

$$|(T_g u)(t)| \leq \int_0^\infty |g(\tau)| \|u\|_\infty\, d\tau = \|g\|_1 \|u\|_\infty.$$

Note also that

$$(T_g e_\lambda)(t) = \int_0^\infty g(\tau) e^{i\lambda(t-\tau)}\, d\tau = (Lg)(i\lambda) e_\lambda(t).$$

Thus T_g maps the space of trigonometric polynomials to itself; hence, by continuity, it maps $AP(\mathbb{R})$ to itself. It still remains to compute the norm.

Without loss of generality, we may assume that $\|g\|_1 = 1$. Given $\epsilon > 0$, let $A > 0$ be chosen such that

$$\int_0^A |g(t)|\, dt > 1 - \epsilon \qquad \text{and} \qquad \int_A^\infty |g(t)|\, dt < \epsilon.$$

We begin by finding a polynomial p such that

$$\int_0^A |g(t) - p(t)|\, dt < \epsilon.$$

This we may certainly do because the continuous functions are dense in the $L^1(0, A)$, and any continuous function is the uniform limit (and hence L^1 limit) of polynomials. Next, we define

$$v(t) = \begin{cases} 0 & \text{if } p(A - t) = 0, \\ |p(A - t)|/p(A - t) & \text{if } p(A - t) \neq 0, \end{cases}$$

so that

$$\int_0^A p(t)v(A - t)\, dt = \int_0^A |p(t)|\, dt > \int_0^A |g(t)|\, dt - \epsilon > 1 - 2\epsilon.$$

Now v is continuous except at any zeroes of p in $[0, A]$. If there are any such zeroes, say $x_1, \ldots, x_N$, then we may modify v by linear interpolation on intervals centred at $x_1, \ldots, x_N$ whose total length is at most $\epsilon/(2\|p\|_\infty)$ to obtain a function $u \in C[0, A]$ such that $\|u\|_\infty \leq 1$, in which case

$$\left| \int_0^A p(t)u(A - t)\, dt \right| > \left| \int_0^A p(t)v(A - t)\, dt \right| - \|p\|_\infty \|v - u\|_1$$
$$\geq (1 - 2\epsilon) - \epsilon = 1 - 3\epsilon.$$

(All norms above apply to functions restricted to $[0, A]$, of course.) Hence

$$\left| \int_0^A g(t)u(A - t)\, dt \right| \geq \left| \int_0^A p(t)u(A - t)\, dt \right| - \int_0^A |p(t) - g(t)|\, dt > 1 - 4\epsilon$$

Finally we may extend u to a continuous $2A$-periodic function with $\|u\|_\infty \leq 1$, in which case

$$\|T_g u\|_\infty \geq \left| \int_0^\infty g(t)u(A - t) \right| > 1 - 5\epsilon,$$

and the result follows.

The situation for $AP_2(\mathbb{R})$ is formally rather simpler. The e_λ form an orthonormal basis of eigenvectors, and the eigenvalues are $(Lg)(i\lambda)$. Hence the operator T_g extends by continuity to a normal operator on the Hilbert space $AP_2(\mathbb{R})$, with norm equal to $\|Lg\|_\infty$. □

As we have seen, almost-periodic functions give an attractive class of persistent signals that behave well in the context of systems theory. In the next section we look at some larger spaces of functions. As will be seen, we lose a little in structure while at the same time gaining in generality.

5.2 Power signal spaces

When we looked at so-called finite-energy signal spaces, there were four natural choices, namely, the discrete-time spaces $\ell^2(\mathbb{Z})$ and $\ell^2(\mathbb{Z}_+)$ and the continuous-time spaces $L^2(\mathbb{R})$ and $L^2(0,\infty)$. A similar problem arises when we look at finite-power spaces: do we use continuous time or discrete time? Do we work on the whole time axis, or just the non-negative half?

Here we shall work with signal spaces in continuous non-negative time only, mentioning the discrete case only briefly. Note first that the following orthogonality formula still holds for $\lambda,\ \mu \in \mathbb{R}$:

$$\lim_{T\to\infty} \frac{1}{T} \int_0^T e^{i\lambda t} e^{-i\lambda\mu}\, dt = \begin{cases} 1 & \text{if } \lambda = \mu, \\ 0 & \text{if } \lambda \neq \mu, \end{cases} \tag{5.13}$$

and hence, if $f(t) = \sum_{k=1}^N a_k e^{i\lambda_k t}$, with $a_1, a_2, \ldots, a_N \in \mathbb{C}$ and distinct real numbers $\lambda_1, \lambda_2, \ldots, \lambda_N$, then

$$\lim_{T\to\infty} \frac{1}{T} \int_0^T |f(t)|^2\, dt = \sum_{k=1}^N |a_k|^2,$$

as in the double-sided case.

By the above remarks, we are motivated to define the *limit power* of a function $f \in L^2_{\text{loc}}(0,\infty)$, when it exists, by the formula

$$\|f\|_P = \left(\lim_{T\to\infty} \frac{1}{T} \int_0^T |f(t)|^2\, dt \right)^{1/2}. \tag{5.14}$$

This formula has been used extensively by engineers but is of limited analytic value because of the following result, given by Mari and others.

Proposition 5.2.1 *The set* **P** *of functions in* L^2_{loc} *for which the limit power exists is not a linear space.*

Proof: We construct two functions $f,\ g \in \mathbf{P}$ such that $f + g \notin \mathbf{P}$. Take $f(t)$ to be equal to 1 for all t and

$$g(t) = (-1)^n \qquad \text{for } 2^{n-1} \leq t < 2^n, \qquad (n \in \mathbb{Z}).$$

It is clear that $\|f\|_P = \|g\|_P = 1$. However, for $M \in \mathbb{N}$,

$$2^{-2M} \int_0^{2^{2M}} |f(t) + g(t)|^2\, dt = 2^{-2M} \sum_{k=-\infty}^M 4 \times 2^{2k-1} = \frac{8}{3},$$

whereas

$$2^{-2M-1} \int_0^{2^{2M+1}} |f(t) + g(t)|^2 \, dt \;\; = \;\; 2^{-2M-1} \sum_{k=-\infty}^{M} 4 \times 2^{2k-1} = \frac{4}{3},$$

and so no limiting value of the power of $f + g$ exists. $\qquad\qquad\qquad\square$

Thus, as it stands, $\| \, . \, \|_P$ is not even a seminorm, let alone a norm, as it is not defined on a linear space. To get round this problem, there are three possible notions of power with which it is convenient to work.

Definition 5.2.2 *The* lim sup power *seminorm is defined for all functions* $f \in L^2_{\mathrm{loc}}$ *for which the quantity exists by the formula*

$$\|f\|_P = \left(\limsup_{T \to \infty} \frac{1}{T} \int_0^T |f(t)|^2 \, dt \right)^{1/2}.$$

The power norms $\|f\|_{\mathcal{B}0}$ *and* $\|f\|_{\mathcal{B}1}$ *are defined by*

$$\|f\|_{\mathcal{B}0} = \left(\sup_{T>0} \frac{1}{T} \int_0^T |f(t)|^2 \, dt \right)^{1/2}$$

and

$$\|f\|_{\mathcal{B}1} = \left(\sup_{T>1} \frac{1}{T} \int_0^T |f(t)|^2 \, dt \right)^{1/2}.$$

The sets of functions in $L^2_{\mathrm{loc}}(0, \infty)$ *for which these quantities are finite are written* $\mathcal{P}$, $\mathcal{B}0$ *and* $\mathcal{B}1$, *respectively.*

We defined $\| \, . \, \|_P$ earlier, by Equation (5.14), and our definition of the $\mathcal{P}$ semi-norm extends the old one to a larger class of functions, which is now a linear space. The $\mathcal{B}1$ norm is sometimes preferred to the $\mathcal{B}0$ norm, since it avoids problems near zero. In particular, we would like an $L^2(0, \infty)$ function to have finite power; however, the example $f(t) = t^{-1/4}\chi_{(0,1)}(t)$ gives, for $0 < T \leq 1$,

$$\frac{1}{T} \int_0^T |f(t)|^2 \, dt = T^{-1} \times (2T^{1/2}) \to \infty$$

as $T \to 0$.

The following proposition collects a few useful facts about these spaces.

Proposition 5.2.3 *The spaces* $\mathcal{P}$, $\mathcal{B}0$ *and* $\mathcal{B}1$ *are linear spaces. Moreover,* $\| \, . \, \|_P$ *is a seminorm and* $\| \, . \, \|_{\mathcal{B}0}$ *and* $\| \, . \, \|_{\mathcal{B}1}$ *are norms. The quantity* $\|f\|_P$ *is finite if and only if* $\|f\|_{\mathcal{B}1}$ *is finite.*

Proof: We leave most of this as an exercise, reminding the reader that one always has the following L^2 triangle inequality:

$$\left(\int_0^T |f(t) + g(t)|^2 \, dt\right)^{1/2} \leq \left(\int_0^T |f(t)|^2 \, dt\right)^{1/2} + \left(\int_0^T |g(t)|^2 \, dt\right)^{1/2}.$$

For the final part, note that if $f \in L^2_{\text{loc}}(0, \infty)$, then the function

$$T \mapsto \frac{1}{T} \int_0^T |f(t)|^2 \, dt$$

is well defined and continuous for $T \geq 1$, and it will have a finite lim sup if and only if it is bounded. $\qquad \square$

In the discrete-time case, we work with a collection of sequences $(a_n)_{n=0}^\infty$ and make the analogous definitions:

$$\|(a_n)\|_{\mathcal{P}} = \left(\limsup_{N \to \infty} \frac{1}{N+1} \sum_{n=0}^N |a_n|^2\right)^{1/2}$$

and

$$\|(a_n)\|_{\mathcal{B}} = \left(\sup_{N \geq 0} \frac{1}{N+1} \sum_{n=0}^N |a_n|^2\right)^{1/2},$$

noting that in this case there is no need to introduce $\mathcal{B}0$ and $\mathcal{B}1$. Life is rather easier here, as the set of sequences for which $\| \, . \, \|_{\mathcal{P}}$ is well defined coincides with the set for which $\| \, . \, \|_{\mathcal{B}}$ is finite.

Our aim now is to show that the bounded causal shift-invariant operators on the power signal spaces defined by $\| \, . \, \|_{\mathcal{P}}$, $\| \, . \, \|_{\mathcal{B}0}$ and $\| \, . \, \|_{\mathcal{B}1}$ are the same as those on $L^2(0, \infty)$ and thus correspond to $H^\infty(\mathbb{C}_+)$ transfer functions, as we saw in Theorem 3.2.3. Thus the methods of H^∞ control can be applied in this context as well.

Note that all the operators in which we are interested, if they are bounded on $L^2(0, \infty)$ or one of the power spaces, must map L^2_{loc} into itself, since if A is a causal operator, then the restriction of Au to an interval $(0, T)$ is always determined by the restriction of u to $(0, T)$. Note that, in the following, whenever we write a norm symbol without a suffix, then for a function it is the L^2 norm, and for an operator it is the operator norm when it acts on L^2.

Theorem 5.2.4 *A causal linear time-invariant operator A satisfies $\|Au\|_{\mathcal{P}} \leq M\|u\|_{\mathcal{P}}$ for all $u \in \mathcal{P}$, for some $M > 0$ independent of u, if and only if A acts as a bounded operator on $L^2(0, \infty)$. Moreover, the induced operator norm $\|A\|_{\mathcal{P}}$ of A acting on $\mathcal{P}$ is the same as its operator norm $\|A\|$ when it is regarded as acting on $L^2(0, \infty)$. The same result holds if we replace $\mathcal{P}$ by $\mathcal{B}0$ or $\mathcal{B}1$.*

Proof:　Suppose that A is bounded on $L^2(0, \infty)$ and $u \in L^2_{\text{loc}}$. Write $y = Au$. Then, since A is causal, by considering $u\chi_{0,T} \in L^2(0, \infty)$ we see that

$$\frac{1}{T} \int_0^T |y(t)|^2 \, dt \leq \|A\|^2 \int_0^T |u(t)|^2 \, dt.$$

By taking the lim sup as $T \to \infty$, we see that $\|Au\|_{\mathcal{P}} \leq \|A\|\|u\|_{\mathcal{P}}$ when this is finite; by taking suprema instead, we obtain the corresponding result for $\mathcal{B}0$ and $\mathcal{B}1$.

The converse is more complicated. Suppose that A is bounded on $\mathcal{P}$ and that $u \in L^2(0, T)$ for some $T > 0$. For each sequence $(\epsilon(t))_{t=0}^{\infty} \subseteq \{-1, 1\}$ of signs, consider the function defined for $t \geq 0$ by

$$u^{\epsilon}(t) = \sum_{k=0}^{\infty} \epsilon(k)u(t - kT).$$

At each point t at most one term in the sum can be non-zero. Observe that $\|u^{\epsilon}\|_{\mathcal{P}} = \|u\|_2 / \sqrt{T}$. Now, looking at each interval of length T in turn, we see that Au has the form

$$(Au)(t) = \sum_{m=0}^{\infty} \sum_{k=0}^{m} \epsilon_k y_{m-k}(t - mT) = \sum_{m=0}^{\infty} z_m(t - mT), \quad \text{say,}$$

for a sequence of functions $(y_k)_{k=0}^{\infty}$ in $L^2(0, T)$. Note that y_0 is just the restriction of Au to $(0, T)$.

Since we always have the triangle inequality $\|v + w\| + \| - v + w\| \geq 2\|v\|$ in any normed space, we may inductively choose a sequence $(\epsilon(t))$ of signs such that the $L^2(0, T)$ norm of each term z_m is at least as great as $\|y_0\|$. Hence

$$\frac{\|Au^{\epsilon}\|_{\mathcal{P}}}{\|u^{\epsilon}\|_{\mathcal{P}}} \geq \frac{\|y_0\|}{\|u\|}.$$

For a fixed u of compact support, we may regard it as lying in $L^2(0, T)$ for sufficiently large T, and thus, by letting $T \to \infty$, we see that $\|Au\|/\|u\| \leq \|A\|_{\mathcal{P}}$; and since the functions of compact support are dense in $L^2(0, \infty)$, we have the required result for $\mathcal{P}$.

Similarly, if A is bounded on $\mathcal{B}0$, then, given $u \in L^2(0, \infty)$ and $\tau > 0$, we have

$$\frac{1}{\tau} \int_0^{\tau} |(Au)(t)|^2 \, dt \leq \|A\|^2_{\mathcal{B}0} \sup_{T>0} \frac{1}{T} \int_0^T |u(t)|^2 \, dt.$$

Now consider the effect of the operator on the shifted function $R_\lambda u$ for $\lambda > 0$. We then have

$$\frac{1}{\tau + \lambda} \int_0^\tau |(Au)(t)|^2 \, dt \;\leq\; \|A\|_{\mathcal{B}0}^2 \sup_{T>0} \frac{1}{T+\lambda} \int_0^T |u(t)|^2 \, dt$$

$$\leq\; \|A\|_{\mathcal{B}0}^2 \frac{1}{\lambda} \|u\|^2.$$

Hence

$$\int_0^\tau |(Au)(t)|^2 \, dt \leq \frac{\tau + \lambda}{\lambda} \|A\|_{\mathcal{B}0}^2 \|u\|^2.$$

Now, letting first $\lambda \to \infty$ and then $\tau \to \infty$, we conclude that $\|Au\|^2 \leq \|A\|_{\mathcal{B}0}^2 \|u\|^2$, as required.

The proof for $\mathcal{B}1$ is almost identical; all that is required is to take $\tau > 1$ and $T > 1$, which makes no difference. $\qquad\square$

Our conclusion is that the operator norm corresponds to the $L^2(0, \infty)$ operator norm, that is, an H^∞ norm, in a variety of situations. In the final section, we shall give another interpretation of this fact under more restrictive hypotheses.

5.3   Spectral distribution functions

We now wish to study the autocorrelation (covariance) function, introduced in (5.7), in more detail. The notation is slightly simpler if we work with functions defined on the whole time-axis, and the autocorrelation of an $L^2_{\mathrm{loc}}(\mathbb{R})$ function f will be defined by

$$\phi_f(x) = [R_{-x}f, f] = \lim_{T \to \infty} \frac{1}{2T} \int_{-T}^T f(x+t)\overline{f(t)} \, dt \qquad (x \in \mathbb{R}),$$

provided that the limit exists. In fact, we are mostly interested in Wiener's class $\mathcal{S}'$, which is the set of functions f for which $\phi_f(x)$ exists for all x, defining a continuous function of x. The following proposition simplifies some of our calculations.

Proposition 5.3.1 *For $f \in \mathcal{S}'$ and for all a, b, $x \in \mathbb{R}$, we have*

$$\phi_f(x) = \lim_{T \to \infty} \frac{1}{2T} \int_{-T+a}^{T+b} f(x+t)\overline{f(t)} \, dt.$$

Proof: Since

$$\lim_{T \to \infty} \frac{1}{2T} \int_{-T-b}^{T+b} |f(t)|^2 \, dt = \phi_f(0) = \lim_{T \to \infty} \frac{1}{2T} \int_{-T}^T |f(t)|^2 \, dt,$$

we see that

$$\lim_{T \to \infty} \frac{1}{2T} \int_T^{T+b} |f(t)|^2 \, dt = 0,$$

and similarly

$$\lim_{T \to \infty} \frac{1}{2T} \int_{-T+a}^{-T} |f(t)|^2 \, dt = 0.$$

A change of variable shows that the same holds when we replace t by $x + t$, and finally the Cauchy–Schwarz inequality implies that

$$\lim_{T \to \infty} \frac{1}{2T} \left(\int_{-T+a}^{-T} + \int_T^{T+b} \right) |f(x + t)||f(t)| \, dt = 0,$$

which gives the required result.   □

Note that the following extension of Proposition 5.2.1 applies, which shows that the methods of linear analysis must be used with caution.

Proposition 5.3.2 *Let f and g be as in Proposition 5.2.1, extended to be iden-tically 0 on the negative time-axis. Then f, $g \in \mathcal{S}'$, but $f + g \notin \mathcal{S}'$.*

Proof: Recall that $f(t) = 1$ for all $t \geq 0$ and

$$g(t) = (-1)^n \qquad \text{for} \quad 2^{n-1} \leq t < 2^n, \qquad (n \in \mathbb{Z}).$$

It is clear that

$$\phi_f(x) = \begin{cases} \lim_{T \to \infty} \frac{1}{2T} \int_0^T dt & \text{if } x \geq 0, \\ \lim_{T \to \infty} \frac{1}{2T} \int_{-x}^T dt & \text{if } x < 0, \end{cases}$$

so that $\phi_f(x) = \frac{1}{2}$ for all x. Moreover, $g(x + t) = g(t)$ unless t lies either in $[-|x|, |x|]$ or in some interval $[2^N - |x|, 2^N + |x|]$ for $2^N \geq |x|$. Thus if $2^k \leq T \leq 2^{k+1}$ and k is large, we have

$$\frac{1}{2T} \int_{-T}^T (g(x + t) - g(t))g(t) \, dt = O(k|x|/2^k),$$

which tends to 0 as $k \to \infty$ and implies that

$$\phi_g(x) = \lim_{T \to \infty} \frac{1}{2T} \int_{-T}^T g(x + t)g(t) \, dt = \phi_g(0) = \frac{1}{2}$$

for all $x \in \mathbb{R}$. Since $\phi_{f+g}(0)$ does not exist, as in Proposition 5.2.1, it is clear that $f + g \notin \mathcal{S}'$.   □

This indicates that the autocorrelation is an inappropriate tool to analyse multi-input systems such as the following:

$$y(t) = u_1(t) + u_2(t),$$

where y is the output and u_1, u_2 are the inputs. Nevertheless, it is still possible to perform some calculations using this concept. Let us begin with some examples.

Example 5.3.3   1. Let $f \in AP(\mathbb{R})$ have a formal Fourier series $\sum_{\lambda \in \mathbb{R}} a_\lambda e_\lambda$, where $a_\lambda = [f, e_\lambda] = 0$ except for countably many λ and $\sum_{\lambda \in \mathbb{R}} |a_\lambda|^2 < \infty$. Then

$$\phi_f(x) = \sum_{\lambda \in \mathbb{R}} |a_\lambda|^2 e^{i\lambda x},$$

which is continuous, since the series converges absolutely and uniformly on $\mathbb{R}$.

2. If $f \in L^2(\mathbb{R})$, then $\phi_f(x) = 0$ for all x. Thus, in general, the autocorrelation does not uniquely determine the function.

3. Probabilists sometimes work with *white noise*, which would ideally be represented by a function f for which $\phi_f(x) = 0$ for all $x \neq 0$, but $\phi_f(0) \neq 0$. There are such functions, for example, $f(t) = e^{it^2}$ (see the exercises), but they do not have the unpredictability one associates with random noise. The following example gives something rather more appropriate.

4. Fix $\delta > 0$ and let $f(t)$ be piecewise constant on intervals $[n\delta, (n+1)\delta)$ for $n \in \mathbb{Z}$, taking values ± 1 independently on each such interval. Then we have $\phi_f(0) = 1$ and $\phi_f(k\delta) = 0$ for all $k \in \mathbb{Z} \setminus \{0\}$, by virtue of the independence (a probabilistic argument shows that the average value of $f(k\delta + t)f(t)$ is 0). For general x we have (if it exists)

$$\phi_f(x) = \lim_{T \to \infty} \frac{1}{2T} \int_{-T}^{T} f(x+t)f(t)\, dt = \lim_{N \to \infty} \frac{1}{2N\delta} \int_{-N\delta}^{N\delta} f(x+t)f(t)\, dt,$$

as is easily seen using the boundedness of f. Thus, for $x \in [k\delta, (k+1)\delta]$ with $k \geq 0$, we have

$$\begin{aligned}
\phi_f(x) &= \lim_{N \to \infty} \frac{1}{2N\delta} \sum_{m=-N}^{N-1} \int_{m\delta}^{(m+1)\delta} f(x+t)f(m\delta)\, dt \\
&= \lim_{N \to \infty} \frac{1}{2N\delta} \sum_{m=-N}^{N-1} \int_{m\delta}^{(m+k+1)\delta-x} f((m+k)\delta)f(m\delta)\, dt \\
&\qquad + \int_{(m+k+1)\delta-x}^{(m+1)\delta} f((m+k+1)\delta)f(m\delta)\, dt \\
&= \lim_{N \to \infty} \frac{1}{2N\delta} \sum_{m=-N}^{N-1} ((k+1)\delta - x)f((m+k)\delta)f(m\delta) \\
&\qquad + (x - k\delta)f((m+k+1)\delta)f(m\delta) \\
&= \frac{((k+1)\delta - x)\phi_f(k\delta) + (x - k\delta)\phi_f((k+1)\delta)}{\delta}.
\end{aligned}$$

A similar calculation holds for $k < 0$, and we conclude that

$$\phi_f(x) = \begin{cases} 1 - \frac{|x|}{\delta} & \text{if } |x| \leq \delta, \\ 0 & \text{otherwise.} \end{cases} \tag{5.15}$$

Let us return to part 1 of Example 5.3.3. In this case one has the following expression for ϕ_f:

$$\phi_f(x) = \frac{1}{2\pi} \int_{-\infty}^{\infty} e^{i\lambda x} d\mu(\lambda), \tag{5.16}$$

where μ is a positive discrete measure with atoms of size $2\pi |[f, e_\lambda]|^2$ at the points where $[f, e_\lambda] \neq 0$. Thus ϕ_f is represented as the inverse Fourier transform of a finite measure. We would like to obtain a similar expression in general, and it is convenient to use the notation of Riemann–Stieltjes integrals.

Definition 5.3.4 *Let* $G : \mathbb{R} \to \mathbb{R}$ *be a bounded monotonically increasing right-continuous function and* $f : \mathbb{R} \to \mathbb{R}$ *a bounded function. Let* $G(\pm\infty)$ *denote the limits of* $f(t)$ *as* $t \to \pm\infty$. *Then the* upper Riemann–Stieltjes sum *of* f *associated with a dissection* $\mathcal{D} = \{x_0, x_1, \ldots, x_n\}$, *where* $x_0 < x_1 < \ldots < x_n$, *is the quantity*

$$S_\mathcal{D}(f) \;=\; (G(x_0) - G(-\infty)) \sup_{t \leq x_0} f(t) + \sum_{j=1}^{n} (G(x_j) - G(x_{j-1})) \sup_{t \in (x_{j-1}, x_j]} f(t)$$
$$+ (G(\infty) - G(x_n)) \sup_{t > x_n} f(t),$$

and the lower Riemann–Stieltjes sum $s_\mathcal{D}(f)$ *is defined similarly, using infima rather than suprema. If* $\sup_\mathcal{D} s_\mathcal{D}(f) = \inf_\mathcal{D} S_\mathcal{D}(f)$, *then their common value* I *is called the* Riemann–Stieltjes integral *of* f *with respect to* G *and written*

$$I = \int_{-\infty}^{\infty} f \, dG.$$

The Riemann–Stieltjes integral has many properties similar to those of the classical Riemann integral, which is a special case, in that

$$\int_a^b f(t) \, dt = \int_{-\infty}^{\infty} f \, dG,$$

on taking

$$G(t) = \begin{cases} 0 & \text{if } t \leq a, \\ t - a & \text{if } a \leq t \leq b, \\ b - a & \text{if } t \geq b. \end{cases}$$

Now formula (5.16) becomes

$$\phi_f(x) = \frac{1}{2\pi} \int_{-\infty}^{\infty} e^{i\lambda x} \, d\sigma_f(\lambda), \tag{5.17}$$

where σ_f is a monotonic increasing step function jumping by an amount $2\pi|[f, e_\lambda]|^2$ at each point λ such that $[f, e_\lambda] \neq 0$. It is common to refer to the function σ_f appearing in (5.17) as the *spectral distribution function* of f. In Wiener's terminology, σ_f is the *spectrum* of f, but this latter term has other meanings and we prefer to avoid it.

Note that (5.17) implies the following formula:

$$\phi_f(0) = \frac{1}{2\pi} \int_{-\infty}^{\infty} d\sigma_f(\lambda) = \frac{\sigma_f(\infty) - \sigma_f(-\infty)}{2\pi}.$$

The fact that, in general, some σ_f exists satisfying (5.17) follows from the following property of ϕ_f, observed by Bochner.

Definition 5.3.5 *A continuous bounded function $\phi : \mathbb{R} \to \mathbb{C}$ is said to be* positive definite *if $\phi(-x) = \overline{\phi(x)}$ for all x and*

$$\sum_{j=1}^{n} \sum_{k=1}^{n} \phi(x_j - x_k) z_j \overline{z_k} \geq 0$$

for all $x_1, \ldots, x_n$ in $\mathbb{R}$ and $z_1, \ldots, z_n$ in $\mathbb{C}$.

The relevance of this concept is explained by the following result.

Proposition 5.3.6 *Suppose that $f \in \mathcal{S}'$. Then ϕ_f is positive definite.*

Proof: By assumption, ϕ_f is continuous, and we also have

$$
\begin{aligned}
|\phi_f(x)| &= \left| \lim_{T \to \infty} \frac{1}{2T} \int_{-T}^{T} f(x+t)\overline{f(t)}\, dt \right| \\
&\leq \left(\lim_{T \to \infty} \frac{1}{2T} \int_{-T}^{T} |f(x+t)|^2\, dt \right)^{1/2} \left(\lim_{T \to \infty} \frac{1}{2T} \int_{-T}^{T} |f(t)|^2\, dt \right)^{1/2} \\
&= \phi_f(0),
\end{aligned}
$$

and so ϕ_f is bounded.

Further,

$$
\begin{aligned}
\phi_f(-x) &= \lim_{T \to \infty} \frac{1}{2T} \int_{-T}^{T} f(-x+t)\overline{f(t)}\, dt \\
&= \lim_{T \to \infty} \frac{1}{2T} \int_{-T-x}^{T-x} f(w)\overline{f(w+x)}\, dw \\
&= \overline{\phi_f(x)}.
\end{aligned}
$$

Also,

$$\sum_{j=1}^{n}\sum_{k=1}^{n}\phi_f(x_j - x_k)z_j\overline{z_k} = \lim_{T\to\infty}\frac{1}{2T}\int_{-T}^{T}\sum_{j=1}^{n}\sum_{k=1}^{n}f(x_j - x_k + t)\overline{f(t)}z_j\overline{z_k}\,dt$$

$$= \lim_{T\to\infty}\frac{1}{2T}\int_{-T}^{T}\sum_{j=1}^{n}f(x_j + w)z_j\sum_{k=1}^{n}\overline{f(x_k + w)\overline{z_k}}\,dw$$

$$= \lim_{T\to\infty}\frac{1}{2T}\int_{-T}^{T}\left|\sum_{j=1}^{n}f(x_j + w)z_j\right|^2 dw \geq 0.$$

$\square$

The same property holds for functions expressed as Fourier integrals of spectral distributions, as in (5.17).

Proposition 5.3.7 *Suppose that $\phi(x) = \frac{1}{2\pi}\int_{-\infty}^{\infty}e^{i\lambda x}\,dG(\lambda)$. Then ϕ is positive definite.*

Proof: The continuity of ϕ follows because

$$|\phi(y) - \phi(x)| \leq \frac{1}{2\pi}\int_{-\infty}^{\infty}|e^{i\lambda y} - e^{i\lambda x}|\,dG(\lambda),$$

and given $\epsilon > 0$ we may find an interval $[a, b]$ such that $G(\infty) - G(b) < \epsilon/6$ and $G(a) - G(-\infty) < \epsilon/6$. Then

$$\int_{-\infty}^{\infty}|e^{i\lambda y} - e^{i\lambda x}|\,dG(\lambda) \leq \frac{\epsilon}{3} + (G(b) - G(a))\sup_{[a,b]}|e^{i\lambda y} - e^{i\lambda x}| + \frac{\epsilon}{3},$$

which is less than ϵ if $|y - x|$ is sufficiently small. The boundedness is proved similarly, but more simply.

Since G is real, it is clear that $\phi(-x) = \overline{\phi(x)}$, and finally

$$\sum_{j=1}^{n}\sum_{k=1}^{n}\phi(x_j - x_k)z_j\overline{z_k} = \int_{-\infty}^{\infty}\left(\sum_{j=1}^{n}z_j e^{i\lambda x_j}\right)\left(\sum_{k=1}^{n}\overline{z_k}e^{-i\lambda x_k}\right)dG(\lambda),$$

which is the integral of a non-negative function and hence is non-negative. $\square$

What is more remarkable is the converse implication.

Theorem 5.3.8 *A bounded continuous function $\phi : \mathbb{R} \to \mathbb{R}$ can be written in the form $\phi(x) = \frac{1}{2\pi}\int_{-\infty}^{\infty}e^{i\lambda x}\,dG(\lambda)$ if and only if it is positive definite. In particular, if $f \in \mathcal{S}'$, then its autocorrelation has a representation*

$$\phi_f(x) = \frac{1}{2\pi}\int_{-\infty}^{\infty}e^{i\lambda x}\,d\sigma_f(\lambda).$$

Proof: We shall sketch the proof, leaving the reader to check any details which do not appear obvious.

Suppose first that $\phi \in L^1(\mathbb{R})$. For a given continuous $g \in L^1(\mathbb{R}) \cap L^2(\mathbb{R})$ form the convolution

$$\Phi(t) = \int_{-\infty}^{\infty} \phi(t + x - y)g(-x)\overline{g(-y)}\, dx\, dy,$$

which is continuous, lies in $L^1(\mathbb{R})$ and satisfies $\Phi(0) \geq 0$, as we see on approximating it by Riemann sums and using the positive definiteness condition.

We now make use several times of Fourier's inversion theorem (see, for example, [69]), which asserts that

$$f(t) = \frac{1}{2\pi} \int_{-\infty}^{\infty} e^{iwt}\widehat{f}(w)\, dw$$

whenever f and $\widehat{f}$ are continuous functions in $L^1(\mathbb{R})$.

By standard properties of Fourier transforms,

$$\widehat{\Phi}(w) = \int_{-\infty}^{\infty} \Phi(t)e^{-iwt}\, dt = \widehat{\phi}(w)\widehat{g}(w)\overline{\widehat{g}(w)}.$$

Since $\widehat{g} \in L^2(\mathbb{R})$, we have $\widehat{\Phi} \in L^1(\mathbb{R})$, so we may conclude that

$$\Phi(0) = \int_{-\infty}^{\infty} \widehat{\phi}(w)|\widehat{g}(w)|^2\, dw \geq 0.$$

It now follows that $\widehat{\phi}(w) \geq 0$ for all w, since g may be chosen so that $\widehat{g}$ is continuous, positive and supported on any small interval that we wish.

Now the conditions that ϕ is continuous, L^1 and bounded, together with $\widehat{\phi} \geq 0$, imply that $\widehat{\phi} \in L^1(\mathbb{R})$. This can be seen by applying Fourier's inversion theorem, for if we take ϕ_m to be the convolution of the functions ϕ and $\frac{m}{\sqrt{\pi}}e^{-m^2 x^2}$ for $m = 1, 2, 3, \ldots$, then

$$\widehat{\phi_m}(w) = \widehat{\phi}(w)e^{-\frac{w^2}{4m^2}},$$

and so

$$\phi_m(0) = \frac{1}{2\pi} \int_{-\infty}^{\infty} \widehat{\phi_m}(w)\, dw.$$

In the limit (by the monotone convergence theorem) we have

$$\phi(0) = \frac{1}{2\pi} \int_{-\infty}^{\infty} \widehat{\phi}(w)\, dw,$$

and so $\hat{\phi} \in L^1(\mathbb{R})$.

Now Fourier's inversion theorem applies in general to ϕ and $\hat{\phi}$, which are both continuous L^1 functions, and so we have the desired result for $\phi \in L^1(\mathbb{R})$.

The remaining part of the proof requires us to approximate ϕ by functions in $L^1(\mathbb{R})$; for example, we may define $\psi_m(t) = e^{-t^2/m^2}\phi(t)$ for $m = 1, 2, 3, \ldots$. This gives us a sequence of expressions involving integrals against measures $\hat{\psi}_m(w)\,dw$, and the proof is accomplished by observing that these have a limit measure μ, such that

$$\phi(x) = \frac{1}{2\pi} \int_{-\infty}^{\infty} e^{i\lambda x}\, d\mu(\lambda) \qquad \text{for all} \quad x \in \mathbb{R}.$$

The last integral can also be expressed as a Riemann–Stieltjes integral by defining $\sigma_f(a) = \mu((-\infty, a])$ for all $a \in \mathbb{R}$.

The final observation follows from Proposition 5.3.6. $\square$

We finish with a result of Wiener, which shows that the action of a convolution operator on a function in $\mathcal{S}'$ will under some circumstances translate into a multiplication operator on the spectral density. Since it is not known what is the best possible result in this direction, we shall just calculate formally and refer the reader to [145] for the rigorous justification, which is surprisingly difficult.

Theorem 5.3.9 *Suppose that $f \in \mathcal{S}'$ and that k is a function on $(0, \infty)$ such that $t \mapsto tk(t)$ lies in $L^1(0, \infty)$ and $t \mapsto (1+t)k(t)$ lies in $L^2(0, \infty)$. Let*

$$g(x) = \int_0^{\infty} k(t)f(x - t)\, dt.$$

Then

$$\sigma_g(\nu) = A + \int_0^{\nu} \left| \int_0^{\infty} k(t)e^{-i\lambda t}\, dt \right|^2 d\sigma_f(\lambda), \qquad \text{for} \quad \nu \in \mathbb{R},$$

where A is a constant. In particular, $\phi_g(0) \leq \|Lk\|_{H^{\infty}(\mathbb{C}_+)}^2 \phi_f(0)$, where L denotes the Laplace transform.

Proof: As explained above, we give just a formal calculation:

$$\begin{aligned}
\phi_g(x) &= \lim_{T\to\infty} \frac{1}{2T} \int_{-T}^{T} g(x+t)\overline{g(t)}\, dt \\
&= \lim_{T\to\infty} \frac{1}{2T} \int_{t=-T}^{T} \left(\int_{u=0}^{\infty} k(u)f(x+t-u)\, du \right) \\
&\qquad\qquad\qquad \times \left(\int_{v=0}^{\infty} \overline{k(v)f(t-v)}\, dv \right) dt
\end{aligned}$$

$$= \int_{u=0}^{\infty} \int_{v=0}^{\infty} k(u)\overline{k(v)}\phi_f(x - u + v) \, dv \, du$$

$$= \frac{1}{2\pi} \int_{u=0}^{\infty} \int_{v=0}^{\infty} \int_{\lambda=-\infty}^{\infty} k(u)\overline{k(v)}e^{i\lambda(x-u+v)} \, d\sigma_f(\lambda) \, dv \, du$$

$$= \frac{1}{2\pi} \int_{\lambda=-\infty}^{\infty} e^{i\lambda x} \left| \int_{u=0}^{\infty} k(u)e^{-i\lambda u} \, du \right|^2 \, d\sigma_f(\lambda),$$

which gives the desired formula. Finally observe that

$$\phi_g(0) = \frac{1}{2\pi} \int_{-\infty}^{\infty} |(Lk)(i\lambda)|^2 \, d\sigma_f(\lambda),$$

$$\leq \|Lk\|_{\infty}^2 \frac{1}{2\pi} \int_{-\infty}^{\infty} d\sigma_f(\lambda) = \|Lk\|_{\infty}^2 \phi_f(0).$$

$\square$

Thus we have finally recovered part of Theorem 5.2.4 by a very roundabout route, having "explained" it in terms of multiplication operators.

Notes

There are many books which treat almost-periodic functions in greater detail, for example [1, 5, 7, 19, 145]. The classical treatments in [5, 7] are probably the closest to the approach given in the text, although modern topological arguments have simplified some of the demonstrations.

Some discussion of linear systems with almost-periodic inputs can be found in [73]. Theorem 5.1.18, at least for $AP_2(\mathbb{R})$, is essentially contained in [59].

Our discussion of power signal spaces and shift-invariant systems is a synthesis of the articles [64, 79, 82, 103].

Mari's result [83] was published in 1996, although it was known earlier (see, for example, [70] and the commentaries by Masani in [144]). In any case, it was necessary to draw it to the attention of the engineering community. Our example is derived from [79].

We draw on [6, 82, 145] for the material of Section 5.3. Proposition 5.3.2 may be new. Other relevant papers in the engineering literature are [95, 152]. The Riemann–Stieltjes integral is presented in [6, 68]. For the spectral analysis of stochastic processes, we refer to [14].

Exercises

1. Verify the orthogonality formulae (5.5) and (5.13).

2. Show directly that the function $f(t) = \cos t + \cos \sqrt{2}t$ is not periodic, by finding all real solutions to the equation $f(t) = 2$ and using the fact that $\sqrt{2}$ is irrational.

3. Suppose that $f \in AP(\mathbb{R})$. By approximating f with trigonometric polynomials, prove that the limit defining $[f, f]$ exists.

4. Give a direct proof of Parseval's identity $[f, f] = \sum_{\lambda \in \mathbb{R}} |[f, e_\lambda]|^2$ for functions $f \in AP(\mathbb{R})$.

5. By evaluating a double integral, prove that, if f_T is a $2T$-periodic function with Fourier coefficients (a_k), then the function g_T given by (5.10) has Fourier coefficients $|a_k|^2$.

6. Use Zorn's lemma to show that any subset $\Lambda \subseteq \mathbb{R}$ contains a maximal subset that is linearly independent over $\mathbb{Q}$.

7. Show that, for a periodic function $g \in C_b(\mathbb{R})$, all the numbers $\lambda \in \mathbb{R}$ for which $[g, e_\lambda] \neq 0$ are integer multiples of some particular real number.

8. A function $f \in C_b(\mathbb{R})$ is said to be *limit-periodic* if it is the uniform limit of a sequence of periodic functions. Assuming the result of Exercise 7, show that f is limit-periodic if and only if it is almost-periodic and all the non-zero numbers $\lambda \in \mathbb{R}$ for which $[f, e_\lambda] \neq 0$ are rational multiples of each other.

9. Suppose that f is a Bohr function for which the numbers λ with $[f, e_\lambda] \neq 0$ are all independent over $\mathbb{Q}$. Use the Fejér–Bochner approximants to show that the Fourier series of f must converge uniformly.

10. Prove in detail the results on $\mathcal{P}$, $\mathcal{B}0$ and $\mathcal{B}1$ given in Proposition 5.2.3.

11. Let $f(t) = e^{i|t|^{1/2}}$. Prove that $\phi_f(x) = 1$ for all x.

12. Let $f(t) = e^{it^2}$. Prove that

$$\phi_f(x) = \begin{cases} 1 & \text{if } x = 0, \\ 0 & \text{if } x \neq 0. \end{cases}$$

Thus ϕ_f exists, but $f \notin \mathcal{S}'$.

13. Calculate the spectral distribution function σ_f in the case that ϕ_f, as given in (5.15), represents an approximation to white noise.

Chapter 6

Delay systems

In this chapter the main aim is to consider an important class of infinite-dimensional systems for which the transfer functions are irrational and finite-dimensional linear algebra no longer provides an appropriate framework in which to proceed. After a brief resumé of the theory of finite-dimensional systems, we give the standard classification theorem for delay systems and then discuss questions of stability, approximation by finite-dimensional systems, and finally stabilization by feedback control.

The methods of this chapter are mostly complex analytical (including the study of poles of meromorphic functions and a certain amount of interpolation theory), but operator-theoretic methods and ideas from approximation theory are also to the fore.

6.1   Background and classification

Finite-dimensional systems can be expressed in matrix terms by the equations

$$
\begin{aligned}
\dot{x}(t) &= Ax(t) + Bu(t), \\
y(t) &= Cx(t) + Du(t), \qquad t \geq 0,
\end{aligned}
\tag{6.1}
$$

where $x(t) \in \mathbb{R}^n$ is the *state* of the system at time t; $u(t) \in \mathbb{R}^m$ denotes the *input* and $y(t) \in \mathbb{R}^p$ is the *output*; and A, B, C, D are matrices of sizes $n \times n$, $n \times m$, $p \times n$ and $p \times m$, respectively. (We could work over $\mathbb{C}$ rather than $\mathbb{R}$, but for physical reasons this is not usually desirable.)

For convenience, and to re-affirm the connection with shift-invariant operators, we shall suppose that $x(0) = 0$, that is, zero initial conditions. Let us write $U = Lu$, the Laplace transform of u, and $Y = Ly$, similarly. Note that the

Laplace transform of $\dot{x}$ is given by

$$
\begin{aligned}
(L\dot{x})(s) &= \int_0^\infty e^{-st}\dot{x}(t)\,dt \\
&= [e^{-st}x(t)]_{t=0}^\infty + s\int_0^\infty x(t)e^{-st}\,dt = s(Lx)(s)
\end{aligned}
$$

for $\mathrm{Re}\,s > 0$. It is then easy to verify that the input–output relation between U and Y is given by $Y(s) = G(s)U(s)$ in some right-hand half-plane, where

$$
G(s) = D + C(sI - A)^{-1}B,
$$

the transfer function, a matrix-valued function with rational entries, since $\det(sI - A)$ is a polynomial in s.

One important example of a finite-dimensional system is given by a differential equation of the form

$$
\sum_{k=0}^n a_k y^{(k)}(t) = \sum_{k=0}^n b_k u^{(k)}(t),
$$

which has a transfer function given by

$$
G(s) = \frac{\sum_{k=0}^n a_k s^k}{\sum_{k=0}^n b_k s^k}.
$$

It is well known, and easy to see, that this can be re-expressed equivalently in the *state space form* (6.1) by working with vectors whose components are $y, \dot{y}, \ddot{y}, \ldots, y^{(n)}$ and $u, \dot{u}, \ddot{u}, \ldots, u^{(n)}$. For example, $\ddot{y} + \omega^2 y = u$ can be replaced by the equivalent set of equations

$$
\frac{d}{dt}\begin{pmatrix} x_1 \\ x_2 \end{pmatrix} = \begin{pmatrix} 0 & 1 \\ -\omega^2 & 0 \end{pmatrix}\begin{pmatrix} x_1 \\ x_2 \end{pmatrix} + \begin{pmatrix} 0 \\ 1 \end{pmatrix}u.
$$

Moving on to infinite-dimensional systems, which are ones for which the state of the system cannot be specified by a finite number of parameters, the simplest class of such systems is the class of delay systems. Many physical systems involve delays: these can be of the order of a fraction of a second for an electrical component to respond, or they can be several hours in the case of complex engineering processes involving the transport of chemicals in pipes. In either case, delays may need to be taken into account.

Let us begin with an example. For calculating transfer functions, note that the Laplace transform of the right shifted function $R_h x : t \mapsto x(t - h)$ is given by

$$
\begin{aligned}
(LR_h x)(s) &= \int_h^\infty e^{-st}x(t - h)\,dt \\
&= \int_0^\infty e^{-s(\tau+h)}x(\tau)\,d\tau, \qquad \text{with } t = \tau + h, \\
&= e^{-sh}(Lx)(s).
\end{aligned}
$$

Example 6.1.1 *The following ideas were presented briefly at the start of Chapter 3, but now we give more details. We begin with the first-order differential equation*

$$\dot{x}(t) + x(t) = u(t),$$

and we take the output equal to the state, that is, $y = x$. By an easy calculation, this system corresponds to the transfer function $G(s) = 1/(s+1)$, which lies in $H^\infty(\mathbb{C}_+)$.

Now suppose that the input arrives after a delay h, that is, the new equation is

$$\dot{x}(t) + x(t) = u(t),$$

and suppose that $u(t) = 0$ on $[-h, 0]$. The new transfer function is then

$$G(s) = \frac{e^{-sh}}{s+1},$$

and this is still in $H^\infty(\mathbb{C}_+)$.

Finally, suppose that the delay occurs in the state, so that we have

$$\dot{x}(t) + x(t-h) = u(t).$$

The transfer function is now

$$G(s) = \frac{1}{s + e^{-sh}},$$

and it can be verified that this is no longer in $H^\infty(\mathbb{C}_+)$ when $h = \pi/2$, as there are then poles of G at $s = \pm i$.

A general way of expressing delay equations in matrix terms is the following (it is not the only possibility):

$$\dot{x}(t) = A_0 x(t) + \sum_{j=1}^{J} A_j x(t - h_j) + B_0 u(t) + \sum_{j=1}^{J} B_j u(t - h_j),$$

$$y(t) = C_0 x(t) + \sum_{j=1}^{J} C_j x(t - h_j) + D_0 u(t) + \sum_{j=1}^{J} D_j u(t - h_j),$$

together with suitable initial conditions on x and u. In this case it can be verified that the transfer function (if it exists) is now a matrix-valued function with meromorphic entries that are in the ring generated by the functions s and e^{-sh_j}, for $j = 1, \ldots, J$.

Thus, whereas the study of finite-dimensional systems may be regarded as the study of rational functions (quotients of polynomials), the study of infinite-dimensional systems may be regarded as the study of quotients of polynomials in s and functions e^{-sh_j}.

There is one elementary case that needs to be discussed before we consider the general question of classifying the transfer functions of delay systems, and that is when G has the form

$$G(s) = \sum_{k=1}^{n} e^{-\lambda_k s} R_k(s),$$

where $\lambda_1, \lambda_2, \ldots, \lambda_n \geq 0$ and each R_k is a rational function of s. For G to lie in $H^{\infty}(\mathbb{C}_+)$ it is clearly sufficient that each R_k be proper (that is, with the degree of the denominator at least as big as the degree of the numerator) and to have no poles in the closed right half-plane. However, degenerate examples such as $G(s) = \frac{1-e^{-s}}{s}$ can occur, so that some of the singularities of the R_k may not be poles of G. In any case, there are only finitely many to check. The fact that each R_k must be proper if G is to lie in $H^{\infty}(\mathbb{C}_+)$ follows easily from the observation that any function $F(s) = \sum_{j=1}^{m} c_j e^{-\mu_j s}$ satisfies

$$\lim_{T \to \infty} \frac{1}{2T} \int_{-T}^{T} |F(i\omega)|^2 \, d\omega = \sum_{j=1}^{m} |c_j|^2,$$

and thus $\limsup_{|\omega| \to \infty} |F(i\omega)| > 0$. (Such results were discussed in greater detail in Section 5.1, in the context of almost-periodic functions.)

A key to the classification of delay systems in the general case, when the number of poles is infinite, lies in the following lemma. (For convenience we assume that the principal value of the argument of a complex number s is chosen to satisfy $-\pi < \arg s \leq \pi$.)

Lemma 6.1.2 *Let $\alpha \in \mathbb{C} \setminus \{0\}$. Then the equation $se^s = \alpha$ has infinitely many solutions, which for large values of $|s|$ have the form $s = x + iy$ with*

$$\begin{aligned} x &= -\log 2n\pi + \log|\alpha| + o(1), \\ y &= \pm 2n\pi \mp \pi/2 + \arg \alpha + o(1), \end{aligned}$$

with n a large positive integer.

Proof: By considering the equations

$$\begin{aligned} x + \log|x + iy| &= \log|\alpha|, \\ y + \arg(x + iy) &= \arg \alpha + 2k\pi, \end{aligned}$$

we see that for any given $\delta > 0$ all solutions to the given equation with n sufficiently large lie in a double sector

$$S_\delta = \{s \in \mathbb{C} : \pi/2 - \delta < |\arg s| < \pi/2 + \delta\}$$

centred on the imaginary axis. Moreover, for R sufficiently large, the mapping $u = s + \log s$ takes $S_{\delta,R} = S_\delta \cap \{s \in \mathbb{C} : |s| > R\}$ bijectively to a region in such a way that points $(r\cos\theta, r\sin\theta)$ are mapped to points $(r\cos\theta + \log r, r\sin\theta + \theta)$ with asymptotically the same arguments as s (or u) tends to infinity.

It is now clear that for large n there are solutions u to $e^u = \alpha$ of the form $u = \log\alpha + 2ik\pi$ or, equivalently, solutions s to $se^s = \alpha$ of the required form, lying in $S_{\delta,R}$. ☐

Note that asymptotic expressions for the solutions of $s^m e^{\lambda s} = \alpha$ for $m \in \mathbb{N}$ and $\lambda \neq 0$ follow immediately, since they are obtained by setting $z = \lambda s/m$ and solving $ze^z = \beta\lambda/m$ for each β with $\beta^m = \alpha$.

We now consider three examples, which illustrate the possible types of behaviour we may encounter.

1. Let $G_1(s) = 1/(1 + e^{-s})$. Then the poles of G_1 lie on the imaginary axis, at the points $(2k+1)\pi i$, $k \in \mathbb{Z}$. More generally, if the poles of a delay system lie in a strip centred on the imaginary axis, then the system is said to be of *neutral type*.

2. Let $G_2(s) = 1/(s - e^{-s})$. Although G_2, like G_1, does not lie in $H^\infty(\mathbb{C}_+)$ (there is a real pole lying between 0 and 1), its poles have the asymptotic form $(-\log 2n\pi, \pm 2n\pi \mp \pi/2)$, by Lemma 6.1.2, and thus there are only finitely many in any right half-plane $\operatorname{Re} s \geq a$. Such a system is said to be of *retarded type*. The first few poles of $G_2(s)$ are shown in Figure 6.1.

3. Let $G_3(s) = 1/(se^{-s} - 1)$. The poles now have the asymptotic form $(\log 2n\pi, \mp 2n\pi \pm \pi/2)$, again by Lemma 6.1.2, and thus there are only finitely many in any left half-plane. Such a system is said to be of *advanced type*.

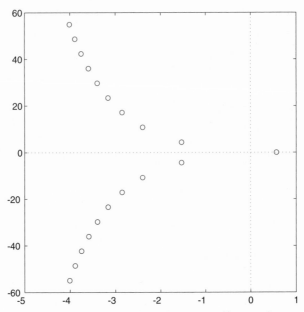

Figure 6.1. Poles of the function $1/(s - e^{-s})$

It is the case that for every delay system with infinitely many poles the poles can be arranged in chains of neutral, retarded or advanced type. In order to determine these chains, we need some notation.

Definition 6.1.3 *Let* $h(s) = \sum_{k=0}^{n} p_k(s)e^{-T_k s}$, *where* $p_0, \ldots, p_n$ *are polynomials of degrees* $d_0, \ldots, d_n$, *with leading coefficients* $c_0, \ldots, c_n$, *and with* $0 = T_0 < \ldots < T_n$. *Then the* distribution diagram *or* Newton diagram *of* h *is the polygonal line joining the points* $P_0 = (T_0, d_0)$, $P_1 = (T_1, d_1), \ldots P_n = (T_n, d_n)$ *with vertices at some of the points* P_k, *which is such that no points* P_k *lie above it (thus it forms a concave polygonal curve).*

For illustration, Figure 6.2 shows the distribution diagram corresponding to the function

$$h(s) = s - 4s^3 e^{-s} + 3s^5 e^{-2s} + 6s^4 e^{-3s} - 5s^5 e^{-4s} + (2s^5 - s^3)e^{-6s} + 2s^3 e^{-7s}. \quad (6.2)$$

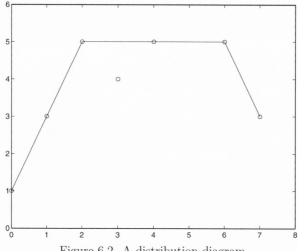

Figure 6.2. A distribution diagram

Theorem 6.1.4 *Let $h(s)$ and its distribution diagram be as given in Definition 6.1.3. Then zeroes of $h(s)$ for large values of $|s|$ are asymptotic to the zeroes of the functions $c_{k(1)}s^{d_{k(1)}}e^{-T_{k(1)}} + \ldots + c_{k(m)}s^{d_{k(m)}}e^{-T_{k(m)}}$, where $P_{k(1)} \ldots P_{k(m)}$ is an edge of the distribution diagram.*

Thus, in the example given in (6.2), to find approximations to the zeroes of large modulus we need only solve each of the following equations independently:

$$s - 4s^3e^{-s} + 3s^5e^{-2s} = 0;$$
$$3s^5e^{-2s} - 5s^5e^{-4s} + 2s^5e^{-6s} = 0;$$
$$2s^5e^{-6s} + 2s^3e^{-7s} = 0.$$

These reduce easily to the equations:

$$s^2e^{-s} = 1 \text{ or } \frac{1}{3};$$
$$e^{-2s} = 1 \text{ or } \frac{3}{2};$$
$$s^2e^s = -1.$$

These in turn can be solved using Lemma 6.1.2, and we see that the first equation gives four advanced chains of zeroes corresponding to $se^{-s/2} = \pm 1, \pm 1/\sqrt{3}$, whereas the second gives two neutral chains, and the third gives two retarded chains, corresponding to $se^{s/2} = \pm i$.

Proof: We sketch the proof. The key lies in the observation that a large zero of h can only arise if two or more terms are of the same order of magnitude and able to cause cancellation. This implies that, at a sufficiently large zero, we have $|s^\lambda e^{-s}|$ bounded away from zero and infinity by absolute constants independent of s, where λ is the gradient of the line segment L passing through two points P_j and P_k.

Now, if P_ℓ is also on the line L, then the term corresponding to P_ℓ is of the same order of magnitude and must be taken into account. If P_ℓ is above the line L, then cancellation between the terms corresponding to P_j and P_k is irrelevant, as the term corresponding to P_ℓ dominates both. Finally, if P_ℓ is below the line, then it corresponds to terms of lower order and is not relevant.

The above argument (which can be written down in more detail, at the expense of totally obscuring the essential idea) shows that the only possible zeroes are asymptotic to the zeroes of various simpler functions, which can be read off the distribution diagram. It remains to show that there really are zeroes near the points specified. This can be done by an argument based on Rouché's theorem: let $s = s_0$ be a zero of the equation $f(s) = c_{k(1)} s^{d_{k(1)}} e^{-T_{k(1)}} + \ldots + c_{k(m)} s^{d_{k(m)}} e^{-T_{k(m)}} = 0$, which corresponds, as above, to the zero of an equation

$$g(s) = \prod_{j=1}^m (s^\lambda e^{-s} - a_j) = 0,$$

with $s_0^\lambda e^{-s_0} = a_r$, say. In the simplest case, there are no repeated a_j, and we find that

$$
\begin{aligned}
g'(s_0) &= (\lambda s_0^{\lambda-1} - s_0^\lambda) e^{-s_0} \prod_{j \neq r} (a_r - a_j) \\
&= -(a_r + O(1/s)) \prod_{j \neq r} (a_r - a_j),
\end{aligned}
$$

with similar bounds on higher derivatives. Note that

$$|g(s_0 + w) - w g'(s_0)| \leq \frac{|w|^2}{2} \sup\{|g''(s_0 + \xi)| : |\xi| \leq |w|\},$$

as may be seen by integration, using the fact that $g(s_0) = 0$. We may therefore conclude that $|h(s) - f(s)| < |f(s)|$ on a small circle centred at s_0 with radius independent of s_0 (for large s_0), and then Rouché's theorem guarantees the existence of a zero inside this circle.

Similar but more complicated arguments can be used in the case of multiple zeroes.   □

The above result enables us to characterize the retarded delay systems, namely, those proper transfer functions that have only finitely many poles in each right half-plane. It is also possible for the denominator of a delay system to be of neutral type while the transfer function is still in $H^\infty(\mathbb{C}_+)$, for example, $G(s) = 1/(2 + e^{-s})$ (see the exercises), but the robust control of such systems is a more delicate matter.

Corollary 6.1.5 *Let $h(s)$ be as in Definition 6.1.3. A necessary and sufficient condition for all the zero chains of h to be of retarded type is that $d_0 > d_k$ for all $k \geq 1$.*

Proof: By Theorem 6.1.4, a necessary and sufficient condition for the presence of only retarded zero chains is that the line segments forming the distribution diagram all have a strictly negative gradient; this is the same as saying that the leftmost point has the largest y-coordinate. □

6.2 Stability

To motivate this section, let us begin with a simple question. For which positive values of h is the transfer function $G_h(s) = \dfrac{1}{s + e^{-sh}}$ in $H^\infty(\mathbb{C}_+)$? An equivalent way of posing this question is to ask when the system

$$\dot{y}(t) + y(t - h) = u(t), \qquad y(0) = 0,$$

is stable, in the sense that its solution defines a bounded shift-invariant operator (from u to y) on $L^2(0, \infty)$.

Since $|G_h(s)| \to 0$ as $|s| \to \infty$ for $s \in \mathbb{C}_+$, it is clear that $G_h \in H^\infty(\mathbb{C}_+)$ if and only if G_h has no poles in the closed right half-plane. In what follows we shall explain how to decide such questions.

Let us begin with the finite-dimensional (rational) case. Suppose that $G(s) = p(s)/q(s)$, where p and q are polynomials (which we suppose to have no common factors – this can be checked using the Euclidean algorithm). Then it is clear that for G to lie in $H^\infty(\mathbb{C}_+)$ it is necessary and sufficient that $\deg p \leq \deg q$ and that q has no zeroes in the closed right half-plane. This latter condition can be checked using the *Routh–Hurwitz test*, which follows.

For physical reasons, we shall work with polynomials over $\mathbb{R}$. It will be convenient to write $q(s) = q_0 s^n + q_1 s^{n-1} + \ldots + q_n$, contrary to the usual convention for labelling coefficients. We say that such a polynomial is *stable* if it has no roots in the closed right half-plane.

One immediate observation, which we shall not use, is that a real stable monic polynomial factors into real quadratic factors $(z-a)(z-\bar{a})$ and real linear factors $(z+b)$, all of whose coefficients are positive; thus, for any stable polynomial, all the coefficients have the same sign. This is not a sufficient condition, and the next result tells us why.

Theorem 6.2.1 (Routh–Hurwitz test) *A real non-constant polynomial*

$$q(s) = q_0 s^n + q_1 s^{n-1} + \ldots + q_n,$$

with $q_0 \neq 0$, is stable if and only if q_0 and q_1 have the same sign (in particular $q_1 \neq 0$), and the degree-$(n-1)$ polynomial

$$r(s) = q(s) - \frac{q_0}{q_1}(q_1 s^n + q_3 s^{n-2} + q_5 s^{n-4} + \ldots)$$

is also stable.

Proof: If we factor q as $q(s) = q_0 \prod_{k=1}^{n}(s - \lambda_k)$, then $q_0 \sum_{k=1}^{n} \lambda_k = -q_1$, which shows immediately that q_0 and q_1 must have the same sign if q is stable. Now consider the family of polynomials

$$r_t(s) = q(s) - t(q_1 s^n + q_3 s^{n-2} + q_5 s^{n-4} + \ldots),$$

where $t \in \mathbb{R}$. Note that $r_0 = q$, and for $t = q_0/q_1$ we have $r_t = r$.

We claim first that all the functions q_t have the same set of imaginary zeroes, including multiplicity. To see this, suppose first that n is even, and write $q = q' + q''$, where q' contains the terms of odd degree and q'' those of even degree. Then

$$r_t = (q'' - tsq') + q',$$

where the first term is an even polynomial and the second is odd. Now, on the imaginary axis the even part takes only real values and the odd part purely imaginary values. Thus, if r_t vanishes at $i\omega$ with multiplicity at least k, then the same is true for its even and odd parts, and thus the same is true for q' and q''. Thus this property holds independently of t. A similar argument holds when n is odd, on decomposing r_t into even and odd parts as

$$r_t = q'' + (q' - tsq'').$$

As we vary t between 0 and q_0/q_1, the zeroes vary continuously and cannot cross the imaginary axis, as we have seen. One zero of $r_t(s) = (q_0 - tq_1)s^n + q_1 s^{n-1} + \ldots$ is asymptotic to $-q_1/(q_0 - tq_1)$, which tends to infinity, and the other zeroes converge to the zeroes of r. The condition that q_0 and q_1 have the same sign is what guarantees that the "lost" zero lies in the right half-plane.   □

Example 6.2.2 Suppose that $q(s) = s^4 + 2s^3 + 3s^2 + 2s + 2$. Then the corresponding polynomial $r(s)$ is given by

$$r(s) = s^4 + 2s^3 + 3s^2 + 2s + 2 - \frac{1}{2}(2s^4 + 2s^2) = 2s^3 + 2s^2 + 2s + 2,$$

and we may repeat the procedure to obtain a new polynomial

$$\tilde{r}(s) = 2s^3 + 2s^2 + 2s + 2 - \frac{2}{2}(2s^3 + 2s) = 2s^2 + 2.$$

This is not a stable polynomial; indeed, its two roots are both on the imaginary axis, which tells us that two of the roots of q are purely imaginary. Indeed, $q(s) = (s^2 + 1)(s^2 + 2s + 2)$.

Consider now the polynomial $q(s) = s^4 + 6s^3 + 15s^2 + 18s + 10$. The Routh–Hurwitz test leads us to

$$r(s) = s^4 + 6s^3 + 15s^2 + 18s + 10 - \frac{1}{6}(6s^4 + 18s^2) = 6s^3 + 12s^2 + 18s + 10,$$

and then to

$$\tilde{r}(s) = 6s^3 + 12s^2 + 18s + 10 - \frac{6}{12}(12s^3 + 10s) = 12s^2 + 13s + 10,$$

and finally (if we wish) to

$$12s^2 + 13s + 10 - \frac{12}{13}(13s^2) = 13s + 10,$$

which is undeniably stable. Indeed, in this case, $q(s) = (s^2 + 2s + 2)(s^2 + 4s + 5)$.

For a cubic polynomial, the first non-trivial case, the Routh–Hurwitz test says that $q_0 s^3 + q_1 s^2 + q_2 s + q_3$ is stable if and only if the coefficients all have the same sign and $q_2 q_1 > q_0 q_3$. We leave the reader to see why this is so. Higher order examples can be analysed by means of a *Routh table*, which is a systematic way of writing down the reduction algorithm that follows by applying Theorem 6.2.1 iteratively.

Let us now discuss the stability of delay systems. We begin with an analysis of functions of the form $P_h(s) = A(s) + B(s)e^{-sh}$, where A and B are polynomials; here, $h \geq 0$ and is allowed to vary. Normally, the most interesting case is when $\deg A > \deg B$ and we are looking at retarded delay systems (most of the zeroes lie in the left half-plane). The object of our discussion is to decide whether such functions have zeroes in the right-hand half-plane. It is easy to see, using Rouché's theorem, that the zeroes, which are infinite in number, vary continuously with h, except at $h = 0$, where only finitely many remain. The important issue, therefore, is to determine where they cross the imaginary axis, and in which direction. A key to the analysis is the following result.

Proposition 6.2.3 *Let $A(s)$ and $B(s)$ be real polynomials. If $P_h(s) = A(s) + B(s)e^{-sh}$ has a zero at a point $s \in i\mathbb{R}$, and $A(s)$ and $B(s)$ are not zero there, then such an s satisfies the equation*

$$A(s)A(-s) = B(s)B(-s). \tag{6.3}$$

Moreover, at such a point s we have

$$\operatorname{sgn} \operatorname{Re} \frac{ds}{dh} = \operatorname{sgn} \operatorname{Re} \frac{1}{s} \left[\frac{B'(s)}{B(s)} - \frac{A'(s)}{A(s)} \right]. \tag{6.4}$$

Proof: From the equation $A(s) + B(s)e^{-sh} = 0$ with $s \in i\mathbb{R}$, we obtain $A(-s) + B(-s)e^{sh} = 0$ by conjugation, and (6.3) follows easily on eliminating the exponential term from the two equations. By elementary calculus we now have

$$(A'(s) + B'(s)e^{-sh} - hB(s)e^{-sh})\frac{ds}{dh} - sB(s)e^{-sh} = 0,$$

which, in conjunction with $e^{-sh} = -A(s)/B(s)$, gives

$$\left[A'(s) - (B'(s) - hB(s))\frac{A(s)}{B(s)} \right] \frac{ds}{dh} = -sA(s),$$

giving

$$\frac{ds}{dh} = -s \left[\frac{A'(s)}{A(s)} - \frac{B'(s)}{B(s)} + h \right]^{-1}.$$

Now $\operatorname{sgn} \operatorname{Re} u = \operatorname{sgn} \operatorname{Re} u^{-1}$ for any $u \in \mathbb{C} \setminus 0$, and h/s is purely imaginary, so the result follows easily.   □

Note that the expression for $\operatorname{sgn} \operatorname{Re} \frac{ds}{dh}$ does not depend on h, which simplifies the discussion. Let us illustrate these ideas by examples.

Example 6.2.4 Consider $P_h(s) = s + e^{-sh}$, which for $h = 0$ has no right half-plane zeroes. Equation (6.3) indicates that imaginary axis zeroes can occur only if $-s^2 = 1$; that is, if $s = \pm i$. It is only necessary to consider one of the conjugate pair, say $s = i$, and solving for h we have

$$i + e^{-ih} = 0, \qquad \text{that is,} \quad h = \frac{\pi}{2} + 2n\pi, \quad n \geq 0.$$

Further we have

$$\operatorname{sgn} \operatorname{Re} \frac{ds}{dh} = \operatorname{sgn} \operatorname{Re} \left(-\frac{1}{s^2} \right) > 0,$$

indicating that zeroes cross from left to right. We may deduce that $P_h(s)$ is stable (has no right half-plane zeroes) if and only if $0 \leq h < \pi/2$.

Example 6.2.5 Consider $P_h(s) = s^3 + s^2 + 2s + 1 + e^{-sh}$. When $h = 0$, this reduces to $(s+1)(s^2+2)$, which is unstable, with zeroes on the imaginary axis. This time Equation (6.3) becomes

$$(s^3 + s^2 + 2s + 1)(-s^3 + s^2 - 2s + 1) = 1$$

or, equivalently,

$$s^2(s^2 + 1)(s^2 + 2) = 0.$$

It is not possible to have a zero at $s = 0$, for any h, so there are two cases remaining.

1. $s = i$: this gives $h = \left(2n + \frac{1}{2}\right)\pi$, with $n \geq 0$. We then calculate (6.4), to obtain

$$-\operatorname{sgn} \operatorname{Re} \frac{3s^2 + 2s + 2}{s(s^3 + s^2 + 2s + 1)} = -\operatorname{sgn} \operatorname{Re} \frac{-1 + 2i}{i \cdot i} < 0,$$

   indicating that the zeroes cross from right to left.

2. $s = \sqrt{2}i$: this gives $h = n\pi\sqrt{2}$ with $n \geq 0$. The expression now becomes

$$-\operatorname{sgn} \operatorname{Re} \frac{-6 + 2\sqrt{2}i + 2}{\sqrt{2}i(-2\sqrt{2}i - 2 + 2\sqrt{2}i + 1)} = -\operatorname{sgn} \frac{2\sqrt{2}i}{-\sqrt{2}i} > 0,$$

   corresponding to a crossing from left to right.

We begin with two unstable zeroes at $h = 0$, moving to the right as h increases. At $h = \pi\sqrt{2}$, they cross back to the left and the system becomes stable immediately afterwards. All is calm until $h = \pi\sqrt{2}$, when two zeroes cross to the right and we lose stability. The next significant value of h is at $5\pi/2$, when there is again a crossing from right to left, resulting in stability. This persists until $h = 2\pi\sqrt{2}$, after which two zeroes cross to the right half-plane once more. From now on, zeroes arrive in the right half-plane faster than they leave it: indeed, there are new arrivals when $h = 3\pi\sqrt{2}$ and no new departure until $h = 9\pi/2$. Accordingly, the stability regions are $\frac{\pi}{2} < h < \pi\sqrt{2}$ and $\frac{5\pi}{2} < h < 2\pi\sqrt{2}$.

We mention two refinements of this method. First, it is possible for $\operatorname{Re} \frac{ds}{dh}$ to vanish at a crossing point, when higher derivatives need to be considered to determine the behaviour of the zeroes.

Second, the method is also applicable to expressions involving more than one delay. Rather than attempt to write down general formulae, which are not very illuminating, let us consider a representative example.

Example 6.2.6 Let $P_h(s) = s + e^{-sh} + e^{-2sh}$, which for $h = 0$ reduces to the stable polynomial $s + 2$. Suppose now that $h > 0$ and that s is a point on the imaginary axis such that

$$s + e^{-sh} + e^{-2sh} = 0, \qquad \text{and hence}$$
$$-s + e^{sh} + e^{2sh} = 0,$$

by complex conjugation. As before, we wish to eliminate the exponential terms from these equations. A simple way to do this is to multiply the second one by e^{-2sh} and eliminate the e^{-2sh} term using the first equation to produce

$$(1 + s^2) + (1 + s)e^{-sh} = 0, \qquad \text{and hence}$$
$$(1 + s^2) + (1 - s)e^{sh} = 0,$$

and finally the polynomial equation

$$(1 + s^2)^2 = (1 + s)(1 - s),$$

which reduces to $s^2(s^2 + 3) = 0$. Again, it is not possible for zeroes to cross the imaginary axis at $s = 0$, so we consider $s = i\sqrt{3}$.

We leave the reader to verify that there are now crossings from left to right whenever $e^{sh} = \frac{1+i\sqrt{3}}{2}$, and so the system is stable for $0 \leq h < \frac{\pi}{3\sqrt{3}}$.

6.3   Rational approximation

The theme of this section is how to approximate a delay system by a finite-dimensional system in an efficient and numerically straightforward way. We shall confine our attentions to delay systems with transfer functions of the form $H(s) = e^{-sT}R(s)$, where $T > 0$ and R is a scalar rational function, and we shall suppose that H lies in $H^\infty(\mathbb{C}_+)$, because we wish to approximate it in the H^∞ norm. It is easy to see that we should therefore make an additional assumption, namely, that H is *strictly proper* (the denominator degree of R exceeds the numerator degree), because otherwise H is not in the closure of the rational functions.

Before beginning a detailed analysis, we make two remarks. First, many unstable systems can be approximated efficiently in the gap topology by taking a coprime factorization and approximating the factors individually, a technique described in [98, 102]. For example, to approximate the system with the transfer function $e^{-s/2}/(s - e^{-s})$, we form the coprime factors $e^{-s/2}/(s + 1)$ and $(s - e^{-s})/(s + 1)$ and then approximate them separately.

Second, more complicated delay systems can be approximated in various ways, for example, by partial fraction expansions [154] (which, however, converge rather slowly, if they do converge at all) or by decomposition techniques as outlined in [45, 46].

One of the keys to the rational approximation of linear systems is the *Hankel operator*. It has many equivalent definitions, so we shall choose one that is well suited to our purposes. (A different, but unitarily equivalent, definition was given in Section 3.4.)

Definition 6.3.1 *Let $H \in L^\infty(i\mathbb{R})$. Then the* Hankel operator $\Gamma = \Gamma_H : H^2(\mathbb{C}_-) \to H^2(\mathbb{C}_+)$ *is defined by*

$$\Gamma U = P_{H^2(\mathbb{C}_+)}(H \cdot U), \qquad U \in H^2(\mathbb{C}_-),$$

where the product $H \cdot U$ is regarded as a function in $L^2(i\mathbb{R})$.

In fact we shall be mostly concerned with the Hankel operators with an analytic symbol $H \in H^2(\mathbb{C}_+)$.

By using the (inverse) bilateral Laplace transform, and writing $U = Lu$ and $H = Lh$, when appropriate, we may obtain a unitarily equivalent form of the operator, namely, $\check{\Gamma} : L^2(-\infty, 0) \to L^2(0, \infty)$, defined by

$$(\check{\Gamma}u)(t) = \int_0^\infty h(t - \tau)u(-\tau)\, d\tau, \qquad t \geq 0, \tag{6.5}$$

at least for $h \in L^1(0, \infty)$ and $u \in L^2(0, \infty)$. This is sometimes regarded as a convolution mapping from past inputs, $u(t)$, $t \leq 0$, to future outputs, $(\check{\Gamma}u)(t)$, $t \geq 0$. We leave this equivalence as an exercise.

The following proposition collects together the results on Hankel operators that we shall need.

Proposition 6.3.2 *Let Γ be the Hankel operator defined in Definition 6.3.1. Then*

- $\|\Gamma\| \leq \|H\|_\infty$;

- *if $H \in H^\infty(\mathbb{C}_+)$ is a rational function of finite degree n, then Γ has rank equal to n (Kronecker's theorem);*

- *if $H \in H^\infty(\mathbb{C}_+)$ is the uniform limit of rational functions, then the operator Γ is compact.*

Proof: The estimate for the norm of Γ is immediate, since it is the composition of a multiplication and a norm-one projection.

Next, if $a \in \mathbb{C}_+$, then, taking $U_a(s) = 1/(s - a)$, we have

$$(P_{H^2(\mathbb{C}_+)}(H \cdot U_a))(s) = \frac{H(s) - H(a)}{s - a},$$

and this has poles at the same places as H, with (at most) the same multiplicities. It therefore lies in the n-dimensional space consisting of rational functions with at most these pre-assigned poles. Thus, since the closed linear span of the functions U_a is all of $H^2(\mathbb{C}_-)$ (as the reader may easily check), we conclude that the rank of Γ is at most n. On the other hand, if $a_1, \ldots, a_m \in \mathbb{C}_+$ satisfy

$$\sum_{k=1}^{m} \lambda_k \frac{H(s) - H(a_k)}{s - a_k} = 0$$

for constants $\lambda_1, \ldots, \lambda_m$ not all zero, and for all $s \in \mathbb{C}_+$, then

$$H(s) \sum_{k=1}^{m} \frac{\lambda_k}{s - a_k} = \sum_{k=1}^{m} \frac{\mu_k}{s - a_k}$$

for some constants $\mu_1, \ldots, \mu_m$, and we conclude that $m \geq n$. Thus the rank of Γ is exactly n.

Finally, if (H_n) is a sequence of rational functions in $H^\infty(\mathbb{C}_+)$ such that $\|H_n - H\|_\infty \to 0$, then the results above show that the corresponding Hankel operators satisfy $\|\Gamma_n - \Gamma\| \to 0$, and so Γ is compact. $\square$

We recall now the singular value decomposition (1.2) of a compact operator $T : \mathcal{H} \to \mathcal{K}$, which in our applications will be taken to be the Hankel operator Γ corresponding to a function H:

$$Tx = \sum_k \sigma_k \langle x, e_k \rangle f_k, \qquad (x \in \mathcal{H}).$$

The following result shows why the singular values are sometimes called *approximation numbers*. Recall that they are arranged in decreasing order.

Lemma 6.3.3 *The singular values* (σ_k) *satisfy*

$$\sigma_{n+1} = \min\{\|T - T'\| : T' : \mathcal{H} \to \mathcal{K}, \ \mathrm{rank}(T') \leq n\}.$$

Proof: Taking $T'x = \sum_{k=1}^{n} \sigma_k \langle x, e_k \rangle f_k$ defines an operator of rank at most k, and $(T - T')x = \sum_{k>n} \sigma_k \langle x, e_k \rangle f_k$, which implies that $\|T - T'\| = \sigma_{n+1}$.

On the other hand, if T' is an arbitrary operator of rank at most n, let us write $\mathcal{L} = \text{lin}\{e_1, \ldots, e_{n+1}\}$. Now $T' : \mathcal{L} \to \mathcal{K}$ is not injective, since $\dim \mathcal{L} > \text{rank}\, T'$, and so we can find a vector $v \in \mathcal{L}$ such that $\|v\| = 1$ and $T'v = 0$. Then $Tv = \sum_{k=1}^{n+1} \sigma_k \langle v, e_k \rangle f_k$, and so

$$\|Tv\|^2 \geq \sum_{k=1}^{n+1} \sigma_k^2 |\langle v, e_k \rangle|^2 \geq \sigma_{n+1}^2 \|v\|^2 = \sigma_{n+1}^2,$$

and hence $\|T - T'\| \geq \sigma_{n+1}$. $\qquad\square$

The following observation is now immediate.

Corollary 6.3.4 *Let* $H \in H^\infty(\mathbb{C}_+)$ *induce a compact Hankel operator* Γ. *Then*

$$\inf\{\|H - R\|_\infty : R \in H^\infty(\mathbb{C}_+),\ R \text{ rational}, \deg R = n\} \geq \sigma_{n+1}(\Gamma).$$

An exact expression for the Hankel singular values (σ_k) is quite complicated in general, even for a function such as $e^{-sT}R(s)$, and involves the solution of a transcendental equation (cf. [45]). Let us do an example, the simplest possible.

Example 6.3.5 Let $H(s) = e^{-sT}/(s + a)$, where a and T are positive real numbers. To calculate the singular values of the Hankel operator, we work with the unitarily equivalent operator $\Gamma' : L^2(0, \infty) \to L^2(0, \infty)$ given by

$$(\Gamma'u)(t) = \int_0^\infty h(t + \tau)u(\tau)\, d\tau,$$

where

$$h(t) = \begin{cases} 0 & \text{for } 0 < T < T, \\ e^{-a(t-T)} & \text{for } t > T. \end{cases}$$

(Note that the Laplace transform of h is just H.) It is easily verified that Γ' is a self-adjoint operator, because h is a real-valued function, and so its singular values are the absolute values of the eigenvalues. The following equation holds for $t < T$:

$$\lambda u(t) = (\Gamma'u)(t) = \int_{T-t}^\infty e^{-a(t-T+\tau)}u(\tau)\, d\tau, \qquad (6.6)$$

which implies that $\lambda \dot{u}(t) = -a\lambda u(t) + u(T - t)$. By differentiating again we soon arrive at the equation

$$\lambda^2 \ddot{u}(t) + (1 - a^2\lambda^2)u(t) = 0.$$

Let ω satisfy $\omega^2 = (1 - a^2\lambda^2)/\lambda^2$, or $\lambda^2 = 1/(a^2 + \omega^2)$. We see that

$$u(t) = Ae^{i\omega t} + Be^{-i\omega t} \qquad (6.7)$$

on $[0, T]$, where A and B are constants. Also, for $t \geq T$,

$$\lambda u(t) = \int_0^\infty e^{-a(t-T+\tau)} u(\tau) \, d\tau = e^{-a(t-T)} \lambda u(T).$$

We now substitute the expression (6.7) for $u(t)$ into (6.6) and equate coefficients of $e^{\pm i\omega t}$. After some manipulation, we find

$$\tan \omega T = -\frac{\omega(3a^2 - \omega^2)}{a(a^2 - 3\omega^2)}, \tag{6.8}$$

and there is an infinite number of solutions, asymptotic to $\omega T = (n + 1/2)\pi$, $n \in \mathbb{Z}$, as can be seen by noting that the right-hand side is approximately $-\omega/(3a)$ for large ω. Thus the Hankel singular values $|\lambda_n|$ are approximately $T/(n\pi)$.

For example, Figure 6.3 shows a plot of the functions $y = \tan \pi \omega$ and $y = -\omega(3 - \omega^2)/(1 - 3\omega^2)$, corresponding to $H(s) = e^{-\pi s}/(s + 1)$.

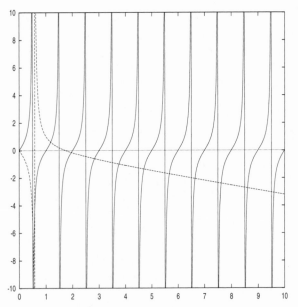

Figure 6.3. Solution of Equation (6.8) with $T = \pi$ and $a = 1$

Although an explicit formula for the singular values is rather complicated, a very transparent asymptotic formula can be given, as follows.

Theorem 6.3.6 *Let $H(s) = e^{-sT}R(s)$ be a strictly proper function in $H^\infty(\mathbb{C}_+)$, where R is a rational function of degree n. Let the set $\{|R(2\pi im/T)| : m \in \mathbb{Z}\}$ be ordered in monotonic decreasing order as $\{\alpha_1, \alpha_2, \ldots\}$. Then the singular values of the associated Hankel operator Γ satisfy*

$$\begin{aligned} \sigma_k &\leq \alpha_{k-2n}, &(k \geq 2n+1), \\ \sigma_k &\geq \alpha_{k+2n}, &(k \geq 1). \end{aligned}$$

Thus if $R(s)s^p \to a \neq 0$ as $|s| \to \infty$, then the singular values satisfy

$$k^p \sigma_k \to |a| \left(\frac{T}{\pi}\right)^p. \tag{6.9}$$

Proof: The method of proof is to show that Γ is a finite-rank perturbation of an operator A whose singular values we already know. We decompose the domain into an orthogonal direct sum $H^2(\mathbb{C}_-) = \Theta H^2(\mathbb{C}_-) \oplus (\Theta H^2(\mathbb{C}_-))^\perp$, where Θ is the inner function given by $\Theta(s) = e^{sT} \in H^\infty(\mathbb{C}_-)$.

For a function $u = \Theta v \in \Theta H^2(\mathbb{C}_-)$, we see that

$$\Gamma u = P_{H^2(\mathbb{C}_+)}\left[e^{-sT}R(s)e^{sT}v(s)\right] = P_{H^2(\mathbb{C}_+)}\left[R(s)v(s)\right] = \Gamma_R v,$$

where Γ_R is the Hankel operator corresponding to R. This therefore lies in the image of Γ_R, an n-dimensional space.

For $m \in \mathbb{Z}$, let $u_m(s)$ be the bilateral Laplace transform of the function $e^{2\pi imt/T}\chi(-T,0)(t)$, namely,

$$u_m(s) = \frac{1 - e^{sT}}{(2\pi im/T) - s}.$$

The (u_m) form an orthogonal basis of $(\Theta H^2(\mathbb{C}_-))^\perp$; indeed, they all have the same norm. Further,

$$\begin{aligned} (\Gamma u_m)(s) &= P_{H^2(\mathbb{C}_+)}\left[R(s)\frac{e^{-sT}-1}{(2\pi im/T) - s}\right] = R(s)u_{-m}(-s) \\ &= R(2\pi im/T)u_{-m}(-s) + \frac{R(s) - R(2\pi im/T)}{(2\pi im/T) - s}(1 - e^{-sT}), \end{aligned}$$

of which the second expression lies in a fixed n-dimensional space, namely, the space of functions $(1 - e^{-sT})Q(s)$, where Q is rational with its set of poles contained in the set of poles of R (including multiplicities).

Thus Γ itself is a finite-rank perturbation of a compact operator A whose singular values are the absolute values of the numbers $|R(2\pi im/T)|$, $m \in \mathbb{Z}$, namely the operator that is zero on $\Theta H^2(\mathbb{C}_-)$ and maps u_m to the function $s \mapsto R(2\pi im/T)u_{-m}(-s)$ in $H^2(\mathbb{C}_+)$.

The perturbation $\Gamma - A$ has rank at most $2n$, and it follows from Corollary 6.3.4 that

$$\sigma(k+2n)(A) \le \sigma_k(\Gamma) \le \sigma_{k-2n}(A),$$

which gives the result.   □

One way of achieving the optimal convergence rate for rational approximants to $e^{-sT}R(s)$ is by means of *Padé approximants*. We shall briefly outline their construction and basic properties.

Definition 6.3.7 *Let F be a function that is analytic in a neighbourhood of zero, with $F(0) \ne 0$. Then an $[n,n]$ Padé approximant to F is a function $G = P_n(s)/Q_n(s)$, where P_n and Q_n are polynomials of degree at most n, such that $Q_n(0) = 1$ and*

$$F(s) = \frac{P_n(s)}{Q_n(s)} + O(s^{2n+1}) \qquad as \quad s \to 0.$$

This is an alternative to truncating the Taylor expansion of F (taking a polynomial approximation), and in many cases it is a better behaved method of approximation. Note that having chosen $Q_n(0) = 1$ we must determine the remaining $2n+1$ coefficients of P_n and Q_n in order to solve the $2n+1$ simultaneous equations implied by the identity

$$F(s)Q_n(s) - P_n(s) = O(s^{2n+1}) \qquad as \quad s \to 0.$$

Our aim now is to calculate the Padé approximants to the function e^{-sT}, where $T > 0$ is fixed. If we take $n = 1$, and consider the functions $P_n(s) = 1 - sT/2$, $Q_n(s) = 1 + sT/2$, then

$$
\begin{aligned}
\frac{P_n(s)}{Q_n(s)} &= (1 - sT/2)(1 - sT/2 + s^2T^2/4 - s^3T^3/8 + \ldots) \\
&= 1 - sT + s^2T^2/2 + \ldots = e^{-sT} + O(s^3),
\end{aligned}
$$

so that we do indeed have a $[1,1]$-Padé approximant. The following theorem gives an expression for the $[n,n]$-approximant to e^s, and it is clear that by replacing s by $-sT$ we obtain the same for e^{-sT}.

Theorem 6.3.8 *For $n \ge 1$, the $[n,n]$-Padé approximant to e^s is given by $G(s) = P(s)/P(-s)$, where*

$$P(s) = \sum_{k=0}^{n} \binom{n}{k} \frac{(2n-k)!}{(2n)!} s^k.$$

Proof: Let f denote the function $f(t) = t^n(1-t)^n$. Then integration by parts gives

$$\int_0^1 e^{ts} f(t)\, dt = [f(t)e^{ts}/s]_0^1 - \frac{1}{s}\int_0^1 e^{ts} f'(t)\, dt$$

$$= -\frac{1}{s}\int_0^1 e^{ts} f'(t)\, dt,$$

and, continuing, we obtain

$$\int_0^1 e^{ts} f(t)\, dt = \frac{(-1)^n}{s^n}\int_0^1 e^{ts} f^{(n)}(t)\, dt.$$

From now on the derivatives of f no longer vanish at the endpoints, and further integration by parts produces

$$\int_0^1 e^{ts} f(t)\, dt = s^{-(2n+1)}\left[e^s(f^{(2n)}(1) - f^{(2n-1)}(1)s + \ldots + (-1)^n f^{(n)}(1)s^n)\right.$$
$$\left. - (f^{(2n)}(0) - f^{(2n-1)}(0)s + \ldots + (-1)^n f^{(n)}(0)s^n)\right],$$

since $f^{(2n+1)}$ is identically zero. Thus

$$e^s(f^{(2n)}(1) - f^{(2n-1)}(1)s + \ldots + (-1)^n f^{(n)}(1)s^n)$$
$$-(f^{(2n)}(0) - f^{(2n-1)}(0)s + \ldots + (-1)^n f^{(n)}(0)s^n) = s^{(2n+1)}\int_0^1 e^{ts} f(t)\, dt.$$

Thus the $[n, n]$-Padé approximant to e^s is $P(s)/Q(s)$, where

$$P(s) = \alpha \sum_{k=0}^n (-1)^k f^{(2n-k)}(0)s^k, \qquad \text{and}$$

$$Q(s) = \alpha \sum_{k=0}^n (-1)^k f^{(2n-k)}(1)s^k,$$

with α chosen to make $Q(0) = 1$, that is,

$$\alpha = 1/f^{(2n)}(1) = (-1)^n/(2n)!.$$

Leibniz's formula gives

$$f^{(r)}(t) = \sum_{m=0}^r {}' \binom{r}{m}\frac{n!}{(n-m)!}t^{n-m}(-1)^{r-m}\frac{n!}{(n-r+m)!}(1-t)^{n-r+m},$$

where the sum only includes terms for which $m \le n$ and $r - m \le n$. We therefore see that for $n \le r \le 2n$ we have

$$f^{(r)}(0) = (-1)^{r-n}\binom{r}{n}n!\frac{n!}{(2n-r)!} = (-1)^{r-n}r!\binom{n}{2n-r}.$$

Thus

$$(-1)^k f^{(2n-k)}(0) = (-1)^n (2n-k)! \binom{n}{k},$$

which gives the formula for $P(s)$. It is easy to see directly that $f^{(m)}(1) = (-1)^m f^{(m)}(0)$ for all m, and thus $Q(s) = P(-s)$, and this completes the proof.
$\square$

Thus, for example, the first three $[n,n]$-Padé approximants to e^{-s} are given by $G_n(s) = P_n(s)/P_n(-s)$, where

$$P_1(s) = 1 - \frac{s}{2},$$

$$P_2(s) = 1 - \frac{s}{2} + \frac{s^2}{12},$$

$$P_3(s) = 1 - \frac{s}{2} + \frac{s^2}{10} - \frac{s^3}{120}.$$

It is easily checked using the Routh–Hurwitz test (Theorem 6.2.1) that these G_n lie in $H^\infty(\mathbb{C}_+)$; indeed, they are inner functions. Such behaviour holds for all the $[n,n]$-Padé approximants to e^{-s} (see, for example, [133]), although we shall not prove this fact. We also assume without proof the following error bound, which can be found in [46].

Theorem 6.3.9 *For $n = 1, 2, \ldots$, let G_n denote the $[n,n]$-Padé approximant to the function e^{-s}. For $s = i\omega$ with $\omega \in \mathbb{R}$, we have*

$$|e^{-i\omega} - G_n(i\omega)| \le \begin{cases} 2\left(\frac{|\omega|}{\psi n}\right)^{2n+1} & \text{for } |\omega| \le \psi n, \\ 2 & \text{for } |\omega| \ge \psi n, \end{cases}$$

where $\psi = 2(\sqrt{2}/e)^{1/2} \approx 1.443$.

Assuming this bound, we can deduce that the Padé method gives an optimal convergence rate for the rational approximation of functions of the form $H(s) = e^{-sT} R(s)$ in $H^\infty(\mathbb{C}_+)$. Note that, for such a function, the low-frequency approximation is taken care of by Theorem 6.3.9, and at high frequencies both the original function and its approximant go to zero.

Theorem 6.3.10 *Suppose that $R \in H_\infty(\mathbb{C}_+)$ and that $M > 0$ and $p \ge 1$ are constants such that $|R(i\omega)| \le M/|\omega|^p$ for all $\omega \in \mathbb{R}$. For $n \ge 1$, let G_n denote the $[n,n]$-Padé approximant to e^{-s}. Then, if $2n + 1 \ge p$, we have*

$$\|e^{-sT} R_n(s) - G_n(sT) R(s)\|_\infty \le 2M \left(\frac{T}{\psi n}\right)^p,$$

where $\psi = 2(\sqrt{2}/e)^{1/2} \approx 1.443$.

Proof: Let $\omega_n = \psi n/T$. Then, by Theorem 6.3.9, for $|\omega| \leq \omega_n$ we have

$$|e^{-i\omega T}R_n(i\omega) - G_n(i\omega T)R(i\omega)| \leq 2 \left(\frac{|\omega|}{\omega_n} \right)^{2n+1} \frac{M}{|\omega|^p} \leq \frac{2M}{\omega_n^p},$$

since $2n + 1 - p \geq 0$, while for $|\omega| \geq \omega_n$ we have

$$|e^{-i\omega T}R_n(i\omega) - G_n(i\omega T)R(i\omega)| \leq \frac{2M}{|\omega|^p} \leq \frac{2M}{\omega_n^p},$$

and hence the result follows. $\qquad\qquad\square$

In practice, there are alternative methods for approximating delay systems, and we now outline a very transparent approach based on shift operators. In Chapter 3 we considered the shift operator $S_\lambda : H^2(\mathbb{C}_+) \rightarrow H^2(\mathbb{C}_+)$ given by multiplication by the function $\Theta_\lambda : s \mapsto e^{-\lambda s}$. For $\lambda > 0$, this is a shift of *infinite multiplicity*, in the sense that $H^2(\mathbb{C}_+) \ominus \Theta_\lambda H^2(\mathbb{C}_+)$ is infinite-dimensional; indeed, it consists of all functions that are Laplace transforms of functions in $L^2(0, \lambda)$. We now think of the approximation procedure as that of approximating S_λ by a shift of finite multiplicity, corresponding to multiplication by a rational inner function.

Let

$$u(s) = f(-s)/f(s),$$

where f is a real polynomial with no zeroes in the closed right half-plane, satisfying $f(0) = 1$. Examples include $f(s) = 1 + s/2$ and $f(s) = 1 + s/2 + s^2/12$, which arise in Padé approximations; another example is the so-called Kautz formula $f(s) = 1 + s/2 + s^2/8$. It is then clear that the function u_n, given by

$$u_n(s) = (u(s/n))^n, \tag{6.10}$$

is inner, and indeed the operator of multiplication by u_n is a shift operator of finite multiplicity (see the exercises). It turns out that the best approximations are obtained when $u(s) \approx e^{-s}$ near $s = 0$, so let $k \geq 1$ denote the index such that, for some constants $A, B, C > 0$, we have

$$A|\omega|^k \leq |e^{-i\omega} - u(i\omega)| \leq B|\omega|^k \qquad \text{for} \quad |\omega| \leq C. \tag{6.11}$$

For the three examples above, we have $k = 3$, $k = 5$ and $k = 3$, respectively. We now have an analogue of Theorem 6.3.9 that holds for u_n.

Lemma 6.3.11 *Under the hypotheses above, we have the error bound*

$$|e^{-i\omega T} - u_n(i\omega T)| \leq \frac{BT^k|\omega|^k}{n^{k-1}} \qquad \text{for} \quad |\omega| \leq nC/T. \tag{6.12}$$

Proof: We write $a^n - b^n = (a - b)(a^{n-1} + a^{n-2}b + \ldots + ab^{n-2} + b^{n-1})$, with $a = e^{-i\omega T/n}$ and $b = u(i\omega T/n)$. Now $|a - b| \leq B|\omega T/n|^k$ for $|\omega T/n| \leq C$ and $|a| = |b| = 1$, which gives the result.   □

It now follows immediately from the dominated convergence theorem that, provided u satisfies (6.11) with $k > 1$, we have strong convergence to S_T of the shift operators associated with u_n. That is,

$$\|u_n(sT)F(s) - e^{-sT}F(s)\|_2 \to 0 \qquad \text{for every} \quad F \in H^2(\mathbb{C}_+).$$

We now analyse the rate of convergence.

Theorem 6.3.12 *Suppose that $R \in H_\infty(\mathbb{C}_+)$, and that $M > 0$ and $k \geq p$ are constants such that (6.11) holds and $|R(i\omega)| \leq M/|\omega|^p$ for all $\omega \in \mathbb{R}$. Let $\omega_n > 0$ satisfy $BT^k\omega_n^k/n^{k-1} = 2$ and suppose that $\omega_n \leq nC/T$. Then*

$$|e^{-sT}R(s) - u_n(sT)R(s)| \leq \frac{2M}{\omega_n^p}.$$

Proof: This follows immediately from (6.12), on considering the following regions:

- $|\omega| \leq \omega_n$, where the bound $BT^k|\omega|^{k-p}M/n^{k-1}$ holds, and

- $|\omega| \geq \omega_n$, where the bound $2M/|\omega|^p$ holds, since we have $|e^{i\omega T} - u_n(i\omega T)| \leq 2$ for all $\omega \in \mathbb{R}$.

In each case the upper bound is at most $BT^k\omega_n^{k-p}M/n^{k-1} = 2M/\omega_p^n$.   □

Thus, for example, if $k = 3$, then ω_n grows as $n^{2/3}$, and the achievable errors for approximating delay systems with $p = 1$, 2 and 3 are, respectively, $O(n^{-2/3})$, $O(n^{-4/3})$ and $O(n^{-2})$. By taking $k = 5$, we have ω_n growing as $n^{4/5}$, and the respective errors for $p = 1, 2, \ldots, 5$ are improved to $O(n^{-4/5})$, $O(n^{-8/5})$, $O(n^{-12/5})$, $O(n^{-16/5})$ and $O(n^{-4})$, which, although not optimal, are quite satisfactory in practice, especially since no complicated computations are required. It can be shown that these error estimates are sharp.

6.4   Stabilization

We saw in Theorem 4.1.8 that the question of finding stabilizing controllers for a linear system P could be reduced to that of constructing a coprime factorization $P = \widetilde{M}^{-1}\widetilde{N} = NM^{-1}$ over $H^\infty(\mathbb{C}_+)$ and finding Bézout factors X, Y, $\widetilde{X}$ and $\widetilde{Y}$ such that

$$\begin{pmatrix} \widetilde{X} & -\widetilde{Y} \\ -\widetilde{N} & \widetilde{M} \end{pmatrix} \begin{pmatrix} M & Y \\ N & X \end{pmatrix} = \begin{pmatrix} M & Y \\ N & X \end{pmatrix} \begin{pmatrix} \widetilde{X} & -\widetilde{Y} \\ -\widetilde{N} & \widetilde{M} \end{pmatrix} = \begin{pmatrix} I & 0 \\ 0 & I \end{pmatrix}. \qquad (6.13)$$

In this case, the set of stabilizing controllers K is parametrized by

$$K = (Y + MQ)(X + NQ)^{-1},$$

and by left coprime factorization

$$K = (\widetilde{X} + Q\widetilde{N})^{-1}(\widetilde{Y} + Q\widetilde{M}),$$

with Q an $H^\infty(\mathbb{C}_+)$ matrix-valued function of the appropriate dimensions. We also saw in Proposition 4.2.11 a constructive algebraic procedure for finding these Bézout factors in the case when P is scalar and rational. We shall now give a somewhat more complicated construction valid for scalar retarded delay systems (where we know that there are at most finitely many unstable poles).

Let us begin with an illustrative example, before writing down more general formulae.

Example 6.4.1 Let $G(s) = \dfrac{e^{-s}}{s - \sigma}$, where $\sigma \geq 0$. For our coprime factorization we take $G = N/M$, where

$$N(s) = \frac{e^{-s}}{s + 1} \quad \text{and} \quad M(s) = \frac{s - \sigma}{s + 1}.$$

In order to have $XM - YN = 1$, we want to find $Y \in H^\infty(\mathbb{C}_+)$ such that $X = M^{-1}(1 + YN)$ also lies in $H^\infty(\mathbb{C}_+)$. Given that M has a zero at σ, we need to ensure that $1 + YN$ also has a zero at σ, that is, that $Y(\sigma)N(\sigma) = -1$. The simplest way to do this is to take $Y(s) = -e^\sigma(\sigma + 1) \in H^\infty(\mathbb{C}_+)$ and then

$$X(s) = \frac{s + 1}{s - \sigma}\left(1 - e^\sigma(\sigma + 1)\frac{e^{-s}}{s + 1}\right) = \frac{s + 1 - (\sigma + 1)e^{\sigma - s}}{s - \sigma} \in H^\infty(\mathbb{C}_+).$$

For the general retarded delay system, the transfer function can be written as $G(s) = h_2(s)/h_1(s)$, where $h_1(s) = \sum_{k=0}^n p_k(s)e^{-T_k s}$ (cf. Definition 6.1.3), with $p_0, \ldots, p_n$ polynomials such that $\deg p_0 > \deg p_k$ for all $k > 0$ and with $0 \leq T_0 < T_1, \ldots < T_n$. The numerator has a similar form, say $h_2(s) = \sum_{j=0}^m q_j(s)e^{-U_j s}$; now the system should be proper ($G(iy)$ should remain bounded as $y \to \pm\infty$), so $\deg q_j \leq \deg p_0$ for each j, and finally $0 \leq U_0 < U_1, \ldots < U_m$. We write $\delta = \deg p_0$. To avoid uninteresting complications, we shall assume that h_1 and h_2 have no common unstable zeroes. If h_1 has no unstable zeroes at all, then h_2/h_1 is already in $H^\infty(\mathbb{C}_+)$, and so we may assume without loss of generality that h_1 has r unstable zeroes with $r \geq 1$.

Theorem 6.4.2 Let h_1 and h_2 be as above. Then a coprime factorization of $G = h_2/h_1$ is obtained by taking $G = N/M$, where

$$M(s) = \frac{h_1(s)}{(s + 1)^\delta} \quad \text{and} \quad N(s) = \frac{h_2(s)}{(s + 1)^\delta}.$$

Corresponding Bézout factors are given by

$$Y(s) = \frac{\mu(s)}{(s+1)^{r-1}} \quad and \quad X(s) = \frac{1 + Y(s)N(s)}{M(s)} = \frac{(s+1)^{\delta} + \frac{\mu(s)h_2(s)}{(s+1)^{r-1}}}{h_1(s)},$$

where μ is a polynomial of degree exactly $r - 1$ chosen such that the function

$$(s+1)^{\delta} + \frac{\mu(s)h_2(s)}{(s+1)^{r-1}}$$

vanishes at every unstable zero of h_1 (and, in the case of multiple zeroes, vanishes according to the appropriate multiplicity).

Proof: There are three things that need to be verified, namely, that X and Y lie in $H^{\infty}(\mathbb{C}_+)$, and that the Bézout identity $XM - YN = 1$ is satisfied. It is obvious that Y is in $H^{\infty}(\mathbb{C}_+)$. Now X is analytic in $\mathbb{C}_+$, because every singularity of h_1 is removed; and since $h_1(s)$ is asymptotic to a non-zero multiple of s^{δ} as $|s| \to \infty$, we see that X is bounded in $\mathbb{C}_+$. Finally, the Bézout identity follows by construction. $\square$

Although algebraic methods can be effective for constructing coprime factorizations and Bézout identities for delay systems (cf. [47]), it seems that they are not always appropriate for finding *normalized* coprime factorizations [104], and approximation techniques are more effective.

Notes

Most books on linear systems and control concentrate on the finite-dimensional case, where the methods required are somewhat more algebraic than analytical. Among books in this area we mention [37, 49, 66, 75, 132, 151].

The classic treatise on delay systems is by Bellman and Cooke [4], on which we have drawn, with some simplifications, for the classification theorem. The asymptotic formulae for poles are used in [154] to construct partial fraction expansions for delay systems, and we also draw on this approach.

The simple proof of the Routh–Hurwitz test is based on the paper of Meinsma [87]. The material on the stability of delay systems is taken mostly from the work of Walton and Marshall [84, 139], although we have also drawn on papers of Chen, Gu and Nett, and Thowsen [17, 129].

The basic properties of Hankel operators may be found in many places, for example [91, 92, 97, 107, 108, 111].

Some approximation techniques are described in [44, 45, 46, 50, 51].

For our discussion of Padé approximants to the exponential function, we refer to [109] for the basic formulae, [119] for more recent analysis, and [46] for detailed error estimates.

The discussion of shift-operator based approximations to the exponential function is based largely on [80, 81].

The formulae in Section 6.4 are based on slightly more general results in [10]. Brethé and Loiseau [13] earlier gave an algorithm for the calculation of Bézout factors in the case when all delays are commensurate. The interpolation method presented here can be generalized to other classes of systems occurring in partial differential equations, such as the heat equation [20, p. 186], and also in the theory of transmission lines [140] (see Exercise 12 below, which is taken from [11]).

Some of the difficulties in studying systems of neutral type are explained in [101].

Exercises

1. Write down determinant equations of the form $\det F(s) = 0$ whose roots are the poles of the transfer functions (from u to y) of the following systems:

    (i) $\dot{y} = Ay + Bz, z = u - Ky$;

    (ii) $\dot{y} = A_0 y + A_1(y - 1) + Bu$.

    Here A, B, K, A_0 and A_1 are matrices of the appropriate sizes.

2. Classify the zero chains of the following:

    (i) $(s + 2) + (4s^2 + 3)e^{-s} + s^3 e^{-4s}$;

    (ii) $(s + 2) + (s + 6)e^{-s} + s^3 e^{-2s}$;

    (iii) $s + se^{-s} - 5e^{-2s}$;

    (iv) $(s^3 + 4s) + (2s + 5)e^{-s} + 3e^{-3s}$.

3. Show that the transfer function $G(s) = \dfrac{1}{e^{-s} + b}$ lies in $H^\infty(\mathbb{C}_+)$ if b is real and $|b| > 1$.

4. Apply the Routh–Hurwitz test to decide which of the following polynomials have all their zeroes in the open left half-plane:

(i) $s^3 + 3s^2 + 4s + 2$;

(ii) $s^4 + 4s^3 + 8s^2 + 8s + 4$;

(iii) $s^4 + 2s^3 + 3s^2 + 38s + 10$.

5. Analyse the following delay expressions to find the range of values of $h \geq 0$ for which they have no zeroes in the closed right half-plane:

(i) $s - 1 + e^{-sh}$;

(ii) $s - \frac{1}{2} + e^{-sh}$;

(iii) $s^2 + 1 + se^{-sh}$.

6. Prove that an $H^\infty(\mathbb{C}_+)$ function $e^{-sT}R(s)$, with $T > 0$ and R rational, lies in the closure of the rational functions if and only if it is strictly proper.

7. Prove the equivalence of the two forms of Hankel operators given in Definition 6.3.1 and (6.5).

8. Calculate the $[4, 4]$-Padé approximant to e^{-s}, and use the Routh–Hurwitz test to show that it is an inner function.

9. Prove that the operator $A_n : H^2(\mathbb{C}_+) \to H^2(\mathbb{C}_+)$ of multiplication by the inner function u_n defined in (6.10) is a shift of finite multiplicity, by calculating $\dim H^2(\mathbb{C}_+) \ominus u_n H^2(\mathbb{C}_+)$.

10. Adapt the proof of Theorem 6.3.12 to show that, if $k \leq p$, then the same method produces rational approximants of error $O(n^{-(k-1)})$.

11. Calculate $H^\infty(\mathbb{C}_+)$ coprime factorizations and some corresponding Bézout factors for the following transfer functions:

(i) $\dfrac{1}{s - e^{-s}}$;

(ii) $\dfrac{e^{-s}}{s^2}$;

(iii) $\dfrac{e^{-s}}{(s - 1)(s - 2)}$.

12. Let $G(s) = \dfrac{\exp(-\sqrt{s})}{s-1}$ (a *fractional system*). Adapt the methods of Section 6.4 to construct an $H^\infty(\mathbb{C}_+)$ coprime factorization and some corresponding Bézout factors for G.

Bibliography

[1] L. Amerio and G. Prouse. *Almost-periodic functions and functional equations.* Van Nostrand Reinhold Co., New York, 1971.

[2] M. Andersson. *Topics in complex analysis.* Springer-Verlag, New York, 1997.

[3] B. Beauzamy. Un opérateur sans sous-espace invariant: simplification de l'exemple de P. Enflo. *Integral Equations Operator Theory*, 8(3):314–384, 1985.

[4] R. Bellman and K. L. Cooke. *Differential-difference equations.* Academic Press, New York, 1963.

[5] A. S. Besicovitch. *Almost periodic functions.* Dover Publications Inc., New York, 1955.

[6] S. Bochner. *Lectures on Fourier integrals. With an author's supplement on monotonic functions, Stieltjes integrals, and harmonic analysis.* Princeton University Press, Princeton, NJ, 1959.

[7] H. Bohr. *Almost periodic functions.* Chelsea Publishing Company, New York, 1947.

[8] B. Bollobás. *Modern graph theory.* Springer-Verlag, New York, 1998.

[9] B. Bollobás. *Linear analysis.* Cambridge University Press, Cambridge, 1999.

[10] C. Bonnet and J. R. Partington. Bézout factors and L^1-optimal controllers for delay systems using a two-parameter compensator scheme. *IEEE Trans. Automat. Control*, 44(8):1512–1521, 1999.

[11] C. Bonnet and J. R. Partington. Coprime factorizations and stability of fractional differential systems. *Systems Control Lett.*, 41(3):167–174, 2000.

[12] C. Bonnet, J. R. Partington, and M. Sorine. Robust control and tracking of a delay system with discontinuous non-linearity in the feedback. *Internat. J. Control*, 72(15):1354–1364, 1999.

[13] D. Brethé and J.-J. Loiseau. Stabilization of linear time-delay systems. *J. Européen Systèmes Automat. (RAIRO-APII-JESA)*, 31(6):1025–1047, 1997.

[14] P. E. Caines. *Linear stochastic systems*. John Wiley & Sons Inc., New York, 1988.

[15] L. Carleson. Interpolations by bounded analytic functions and the corona problem. *Ann. of Math. (2)*, 76:547–559, 1962.

[16] I. Chalendar and J. Esterle. Le problème du sous-espace invariant. In *Development of mathematics 1950–2000*, pages 235–267. Birkhäuser Verlag, Basel, 2000.

[17] J. Chen, G. Gu, and C. N. Nett. A new method for computing delay margins for stability of linear delay systems. *Systems Control Lett.*, 26(2):107–117, 1995.

[18] K. F. Clancey and I. Gohberg. *Factorization of matrix functions and singular integral operators*, volume 3 of *Operator Theory: Advances and Applications*. Birkhäuser Verlag, Basel, 1981.

[19] C. Corduneanu. *Almost periodic functions*. Interscience Publishers [John Wiley & Sons], New York – London – Sydney, 1968.

[20] R. F. Curtain and H. Zwart. *An introduction to infinite-dimensional linear systems theory*. Springer-Verlag, New York, 1995.

[21] H. G. Dales. *Banach algebras and automatic continuity*. The Clarendon Press, Oxford University Press, New York, 2000.

[22] E. B. Davies. *One-parameter semigroups*. Academic Press Inc., London, 1980.

[23] E. B. Davies. *Heat kernels and spectral theory*. Cambridge University Press, Cambridge, 1990.

[24] R. G. Douglas. *Banach algebra techniques in operator theory*, 2nd edition. Springer-Verlag, New York, 1998.

[25] N. Dunford and J. T. Schwartz. *Linear operators. Part I*. John Wiley & Sons Inc., New York, 1988.

[26] P. L. Duren. *Theory of H^p spaces*, 2nd edition. Dover Publications Inc., New York, 2000.

[27] A. K. El-Sakkary. The gap metric: robustness of stabilization of feedback systems. *IEEE Trans. Automat. Control*, 30(3):240–247, 1985.

[28] A. K. El-Sakkary. Estimating robustness on the Riemann sphere. *Internat. J. Control*, 49(2):561–567, 1989.

[29] P. Enflo. On the invariant subspace problem for Banach spaces. *Acta Math.*, 158(3-4):213–313, 1987.

[30] K.-J. Engel and R. Nagel. *One-parameter semigroups for linear evolution equations*, volume 194 of Graduate Texts in Mathematics. Springer-Verlag, New York, 2000.

[31] A. Feintuch. *Robust control theory in Hilbert space*. Springer-Verlag, New York, 1998.

[32] A. Feintuch and R. Saeks. *System theory, A Hilbert space approach*. Academic Press Inc., New York, 1982.

[33] C. Foias and A. E. Frazho. *The commutant lifting approach to interpolation problems*. Birkhäuser Verlag, Basel, 1990.

[34] C. Foias, A. E. Frazho, I. Gohberg, and M. A. Kaashoek. *Metric constrained interpolation, commutant lifting and systems*. Birkhäuser Verlag, Basel, 1998.

[35] C. Foias, T. T. Georgiou, and M. C. Smith. Robust stability of feedback systems: a geometric approach using the gap metric. *SIAM J. Control Optim.*, 31(6):1518–1537, 1993.

[36] Y. Fourès and I. E. Segal. Causality and analyticity. *Trans. Amer. Math. Soc.*, 78:385–405, 1955.

[37] B. A. Francis. *A course in H_∞ control theory*, volume 88 of Lecture Notes in Control and Information Sciences. Springer-Verlag, Berlin, 1987.

[38] P. A. Fuhrmann. *Linear systems and operators in Hilbert space*. McGraw-Hill International Book Co., New York, 1981.

[39] J. B. Garnett. *Bounded analytic functions*. Academic Press, New-York, 1981.

[40] T. T. Georgiou. On the computation of the gap metric. *Systems Control Lett.*, 11(4):253–257, 1988.

[41] T. T. Georgiou and M. C. Smith. Optimal robustness in the gap metric. *IEEE Trans. Automat. Control*, 35(6):673–686, 1990.

[42] T. T. Georgiou and M. C. Smith. Graphs, causality, and stabilizability: linear, shift-invariant systems on $\mathcal{L}_2[0, \infty)$. *Math. Control Signals Systems*, 6(3):195–223, 1993.

[43] T. T. Georgiou and M. C. Smith. Intrinsic difficulties in using the doubly-infinite time axis for input-output control theory. *IEEE Trans. Automat. Control*, 40(3):516–518, 1995.

[44] K. Glover, R. F. Curtain, and J. R. Partington. Realisation and approximation of linear infinite-dimensional systems with error bounds. *SIAM J. Control Optim.*, 26(4):863–898, 1988.

[45] K. Glover, J. Lam, and J. R. Partington. Rational approximation of a class of infinite-dimensional systems. I. Singular values of Hankel operators. *Math. Control Signals Systems*, 3(4):325–344, 1990.

[46] K. Glover, J. Lam, and J. R. Partington. Rational approximation of a class of infinite-dimensional systems. II. Optimal convergence rates of L_∞ approximants. *Math. Control Signals Systems*, 4(3):233–246, 1991.

[47] H. Gluesing-Luerssen. *Linear delay-differential systems with commensurate delays: an algebraic approach*, volume 1770 of Lecture Notes in Mathematics. Springer-Verlag, Berlin, 2002.

[48] I. Gohberg and S. Goldberg. *Basic operator theory*. Birkhäuser Verlag, Boston, MA, 1981.

[49] M. Green and D. J. N. Limebeer. *Linear robust control*. Prentice-Hall Inc., Englewood Cliffs, NJ, 1995.

[50] G. Gu, P. P. Khargonekar, and E. B. Lee. Approximation of infinite-dimensional systems. *IEEE Trans. Automat. Control*, 34(6):610–618, 1989.

[51] G. Gu, P. P. Khargonekar, E. B. Lee, and P. Misra. Finite-dimensional approximations of unstable infinite-dimensional systems. *SIAM J. Control Optim.*, 30(3):704–716, 1992.

[52] C. J. Harris and J. M. E. Valença. *The stability of input-output dynamical systems*. Academic Press Inc., London, 1983.

[53] P. Hartman and A. Wintner. The spectra of Toeplitz's matrices. *Amer. J. Math.*, 76:867–882, 1954.

[54] W. K. Hayman. *Meromorphic functions*. Oxford Mathematical Monographs. Clarendon Press, Oxford, 1964.

[55] H. Helson. *Lectures on invariant subspaces*. Academic Press, New York – London, 1964.

[56] E. Hille and R. S. Phillips. *Functional analysis and semi-groups*. American Mathematical Society, Providence, RI, 1957.

[57] K. Hoffman. *Banach spaces of analytic functions.* Prentice-Hall Inc., Englewood Cliffs, NJ, 1965.

[58] B. Jacob. An operator theoretical approach towards systems over the signal space $\ell_2(\mathbb{Z})$. *Integral Equations Operator Theory,* 46:189–214, 2003.

[59] B. Jacob, M. Larsen, and H. Zwart. Corrections and extensions of: "Optimal control of linear systems with almost periodic inputs" by G. Da Prato and A. Ichikawa. *SIAM J. Control Optim.,* 36(4):1473–1480, 1998.

[60] B. Jacob and J. R. Partington. Admissibility of control and observation operators for semigroups: a survey. *Proceedings of IWOTA-2002,* Birkhäuser Verlag, to appear.

[61] B. Jacob and J. R. Partington. Graphs, closability, and causality of linear time-invariant discrete-time systems. *Internat. J. Control,* 73(11):1051–1060, 2000.

[62] B. Jacob and J. R. Partington. On the boundedness and continuity of the spectral factorization mapping. *SIAM J. Control Optim.,* 40(1):88–106, 2001.

[63] B. Jacob and J. R. Partington. The Weiss conjecture on admissibility of observation operators for contraction semigroups. *Integral Equations Operator Theory,* 40(2):231–243, 2001.

[64] B. Jacob, J. R. Partington, and B. Ünalmiş. On discrete-time linear systems with almost-periodic kernels. *J. Math. Anal. Appl.,* 252(2):549–570, 2000.

[65] B. Jacob and H. Zwart. Counterexamples concerning observation operators for C_0-semigroups. *SIAM J. Control Optim.,* to appear.

[66] T. Kailath. *Linear systems.* Prentice-Hall Inc., Englewood Cliffs, NJ, 1980.

[67] T. Kato. *Perturbation theory for linear operators.* Springer-Verlag, Berlin, 1995.

[68] A. N. Kolmogorov and S. V. Fomīn. *Introductory real analysis.* Dover Publications Inc., New York, 1975. Translated from the second Russian edition and edited by Richard A. Silverman.

[69] T. W. Körner. *Fourier analysis.* Cambridge University Press, Cambridge, 1988.

[70] K. S. Lau and J. K. Lee. On generalized harmonic analysis. *Trans. Amer. Math. Soc.,* 259(1):75–97, 1980.

[71] K. B. Laursen and M. M. Neumann. *An introduction to local spectral theory.* The Clarendon Press, Oxford University Press, New York, 2000.

[72] P. D. Lax. Translation invariant spaces. *Acta Math.*, 101:163–178, 1959.

[73] B. M. Levitan and V. V. Zhikov. *Almost periodic functions and differential equations.* Cambridge University Press, Cambridge, 1982.

[74] R. J. Loy. Continuity of linear operators commuting with shifts. *J. Functional Analysis*, 17:48–60, 1974.

[75] J. M. Maciejowski. *Multivariable feedback design.* Addison-Wesley, Wokingham, Berkshire, 1989.

[76] P. M. Mäkilä. On three puzzles in robust control. *IEEE Trans. Automat. Control*, 45(3):552–556, 2000.

[77] P. M. Mäkilä. When is a linear convolution system stabilizable? *Systems Control Lett.*, 46:371–378, 2002.

[78] P. M. Mäkilä and J. R. Partington. Robust stabilization—BIBO stability, distance notions and robustness optimization. *Automatica J. IFAC*, 29(3):681–693, 1993.

[79] P. M. Mäkilä and J. R. Partington. Lethargy results in LTI system modelling. *Automatica*, 34:1061–1070, 1998.

[80] P. M. Mäkilä and J. R. Partington. Laguerre and Kautz shift approximations of delay systems. *Internat. J. Control*, 72(10):932–946, 1999.

[81] P. M. Mäkilä and J. R. Partington. Shift operator induced approximations of delay systems. *SIAM J. Control Optim.*, 37(6):1897–1912, 1999.

[82] P. M. Mäkilä, J. R. Partington, and T. Norlander. Bounded power signal spaces for robust control and modeling. *SIAM J. Control Optim.*, 37(1):92–117, 1999.

[83] J. Mari. A counterexample in power signals space. *IEEE Trans. Automat. Control*, 41(1):115–116, 1996.

[84] J.E. Marshall, H. Górecki, K. Walton, and A. Korytowski. *Time-delay systems: stability and performance criteria with applications.* Ellis Horwood, London, 1992.

[85] L. Máté. Automatic continuity of shift-invariant linear operators. *Vikram Math. J.*, 4:15–23, 1983.

[86] D. C. McFarlane and K. Glover. *Robust controller design using normalized coprime factor plant descriptions.* Springer-Verlag, Berlin, 1990.

[87] G. Meinsma. Elementary proof of the Routh-Hurwitz test. *Systems Control Lett.*, 25(4):237–242, 1995.

[88] G. Meinsma and H. Zwart. On $\mathcal{H}_\infty$ control for dead-time systems. *IEEE Trans. Automat. Control*, 45(2):272–285, 2000.

[89] Z. Nehari. On bounded bilinear forms. *Ann. of Math. (2)*, 65:153–162, 1957.

[90] N. K. Nikolski. *Invitation aux techniques des espaces de Hardy.* University of Bordeaux, 1992. Lecture notes.

[91] N. K. Nikolski. *Operators, functions, and systems: an easy reading. Vol. 1*, volume 92 of Mathematical Surveys and Monographs. American Mathematical Society, Providence, RI, 2002.

[92] N. K. Nikol'skiĭ. *Treatise on the shift operator*, volume 273 of Grundlehren der mathematischen Wissenschafte. Springer-Verlag, Berlin, 1986.

[93] E. Nordgren, P. Rosenthal, and F. S. Wintrobe. Invertible composition operators on H^p. *J. Funct. Anal.*, 73(2):324–344, 1987.

[94] R. J. Ober and J. A. Sefton. Stability of control systems and graphs of linear systems. *Systems Control Lett.*, 17(4):265–280, 1991.

[95] F. Paganini. A set-based approach for white noise modeling. *IEEE Trans. Automat. Control*, 41(10):1453–1465, 1996.

[96] L. B. Page. Bounded and compact vectorial Hankel operators. *Trans. Amer. Math. Soc.*, 150:529–539, 1970.

[97] J. R. Partington. *An introduction to Hankel operators.* Cambridge University Press, Cambridge, 1988.

[98] J. R. Partington. Approximation of unstable infinite-dimensional systems using coprime factors. *Systems Control Lett.*, 16(2):89–96, 1991.

[99] J. R. Partington. Robust control and approximation in the chordal metric. In *Robust control (Tokyo, 1991)*, pages 82–89. Springer-Verlag, Berlin, 1992.

[100] J. R. Partington. *Interpolation, identification, and sampling.* The Clarendon Press, Oxford University Press, Oxford, 1997.

[101] J. R. Partington and C. Bonnet. H_∞ and BIBO stabilization of delay systems of neutral type. *Systems Control Lett.*, to appear.

[102] J. R. Partington and K. Glover. Robust stabilization of delay systems by approximation of coprime factors. *Systems Control Lett.*, 14(4):325–331, 1990.

[103] J. R. Partington and P. M. Mäkilä. On system gains for linear and nonlinear systems. *Systems Control Lett.*, 46:129–136, 2002.

[104] J. R. Partington and G. K. Sankaran. Algebraic construction of normalized coprime factors for delay systems. *Math. Control Signals Systems*, 15(1):1–12, 2002.

[105] J. R. Partington and B. Ünalmış. On the representation of shift-invariant operators by transfer functions. *Systems Control Lett.*, 33(1):25–30, 1998.

[106] A. Pazy. *Semigroups of linear operators and applications to partial differential equations.* Springer-Verlag, New York, 1983.

[107] V. V. Peller. An excursion into the theory of Hankel operators. In *Holomorphic spaces (Berkeley, CA, 1995)*, pages 65–120. Cambridge University Press, Cambridge, 1998.

[108] V. V. Peller. *Hankel operators and their applications.* Springer Monographs in Mathematics. Springer-Verlag, New York, 2003.

[109] O. Perron. *Die Lehre von den Kettenbrüchen.* B. G. Teubner, Leipzig, 1913.

[110] J. W. Polderman and J. C. Willems. *Introduction to mathematical systems theory, a behavioral approach.* Springer-Verlag, New York, 1998.

[111] S. C. Power. *Hankel operators on Hilbert space.* Pitman (Advanced Publishing Program), Boston, MA, 1982.

[112] L. Qiu and E. J. Davison. Pointwise gap metrics on transfer matrices. *IEEE Trans. Automat. Control*, 37(6):741–758, 1992.

[113] C. J. Read. A solution to the invariant subspace problem. *Bull. London Math. Soc.*, 16(4):337–401, 1984.

[114] C. J. Read. A solution to the invariant subspace problem on the space l_1. *Bull. London Math. Soc.*, 17(4):305–317, 1985.

[115] M. Rosenblum and J. Rovnyak. *Hardy classes and operator theory.* Dover Publications Inc., Mineola, NY, 1997.

[116] W. Rudin. *Real and complex analysis.* McGraw-Hill, New York, 1986.

[117] W. Rudin. *Functional analysis.* McGraw-Hill Inc., New York, 1991.

[118] B. P. Rynne and M. A. Youngson. *Linear functional analysis.* Springer-Verlag London Ltd., London, 2000.

[119] E. B. Saff and R. S. Varga. On the zeros and poles of Padé approximants to e^z. *Numer. Math.*, 25(1):1–14, 1975/76.

[120] D. Sarason. Generalized interpolation in H^∞. *Trans. Amer. Math. Soc.*, 127:179–203, 1967.

[121] J. A. Sefton and R. J. Ober. Graphs of linear systems and stabilization. In *Recent advances in mathematical theory of systems, control, networks and signal processing, I (Kobe, 1991)*, pages 351–356. Mita, Tokyo, 1992.

[122] J. A. Sefton and R. J. Ober. On the gap metric and coprime factor perturbations. *Automatica J. IFAC*, 29(3):723–734, 1993.

[123] A. M. Sinclair. *Automatic continuity of linear operators,* no. 21 of London Mathematical Society Lecture Note Series. Cambridge University Press, Cambridge, 1976.

[124] M. C. Smith. On stabilization and the existence of coprime factorizations. *IEEE Trans. Automat. Control*, 34(9):1005–1007, 1989.

[125] T. P. Srinivasan. Simply invariant subspaces. *Bull. Amer. Math. Soc.*, 69:706–709, 1963.

[126] T. P. Srinivasan. Doubly invariant subspaces. *Pacific J. Math.*, 14:701–707, 1964.

[127] B. Sz.-Nagy and C. Foiaş. Dilatation des commutants d'opérateurs. *C. R. Acad. Sci. Paris Sér. A-B*, 266:A493–A495, 1968.

[128] B. Sz.-Nagy and C. Foiaş. *Harmonic analysis of operators on Hilbert space.* North-Holland Publishing Co., Amsterdam, 1970.

[129] A. Thowsen. An analytic stability test for a class of time-delay systems. *IEEE Trans. Automat. Control*, 26(3):735–736, 1981.

[130] S. Treil. The stable rank of the algebra H^∞ equals 1. *J. Funct. Anal.*, 109(1):130–154, 1992.

[131] S. Treil. A counterexample on continuous coprime factors. *IEEE Trans. Automat. Control*, 39(6):1262–1263, 1994.

[132] H. L. Trentelman, A. A. Stoorvogel, and M. Hautus. *Control theory for linear systems.* Springer-Verlag London Ltd., London, 2001.

[133] H. van Rossum. On the poles of Padé approximations to e^z. *Nieuw Arch. Wisk. (3)*, 19:37–45, 1971.

[134] M. Vidyasagar. The graph metric for unstable plants and robustness estimates for feedback stability. *IEEE Trans. Automat. Control*, 29(5):403–418, 1984.

[135] M. Vidyasagar. *Control system synthesis, a factorization approach.* MIT Press, Cambridge, MA, 1985.

[136] M. Vidyasagar and H. Kimura. Robust controllers for uncertain linear systems. *Automatica*, 22:85–94, 1986.

[137] M. Vidyasagar, H. Schneider, and B. A. Francis. Algebraic and topological aspects of feedback stabilization. *IEEE Trans. Automat. Control*, 27(4):880–894, 1982.

[138] G. Vinnicombe. Frequency domain uncertainty and the graph topology. *IEEE Trans. Automat. Control*, 38(9):1371–1383, 1993.

[139] K. Walton and J. E. Marshall. Direct method for TDS stability analysis. *IEE proceedings D, control theory and applications*, 134:101–107, 1987.

[140] E. Weber. *Linear transient analysis. Volume II.* John Wiley and Sons, Inc., New York, 1956.

[141] G. Weiss. Admissible observation operators for linear semigroups. *Israel J. Math.*, 65(1):17–43, 1989.

[142] G. Weiss. Representation of shift-invariant operators on L^2 by H^∞ transfer functions: an elementary proof, a generalization to L^p, and a counterexample for L^∞. *Math. Control Signals Systems*, 4(2):193–203, 1991.

[143] G. Weiss. A powerful generalization of the Carleson measure theorem? In *Open problems in mathematical systems and control theory*, pages 267–272. Springer-Verlag London Inc., London, 1999.

[144] N. Wiener. *Collected works with commentaries. Vol. II*, volume 15 of Mathematicians of Our Time. MIT Press, Cambridge, Mass., 1979. Generalized harmonic analysis and Tauberian theory; classical harmonic and complex analysis, Edited by P. Masani.

[145] N. Wiener. *The Fourier integral and certain of its applications.* Cambridge University Press, Cambridge, 1988.

[146] K. Yosida. *Functional analysis.* Springer-Verlag, Berlin, 1995.

[147] D. C. Youla, J. J. Bongiorno, Jr., and H. A. Jabr. Modern Wiener-Hopf design of optimal controllers. I. The single-input-output case. *IEEE Trans. Automatic Control*, AC-21(1):3–13, 1976.

[148] D. C. Youla, H. A. Jabr, and J. J. Bongiorno, Jr. Modern Wiener-Hopf design of optimal controllers. II. The multivariable case. *IEEE Trans. Automatic Control*, AC-21(3):319–338, 1976.

[149] N. Young. *An introduction to Hilbert space*. Cambridge University Press, Cambridge, 1988.

[150] G. Zames and A. El-Sakkary. Uncertainty in unstable systems: the gap metric. In *Control science and technology for the progress of society, Vol. 1 (Kyoto, 1981)*, pages 149–152. IFAC, Laxenburg, 1982.

[151] K. Zhou, J. C. Doyle, and K. Glover. *Robust and optimal control*. Prentice Hall, Upper Saddle River, NJ, 1996.

[152] K. Zhou, K. Glover, B. Bodenheimer, and J. Doyle. Mixed $\mathcal{H}_2$ and $\mathcal{H}_\infty$ performance objectives. I. Robust performance analysis. *IEEE Trans. Automat. Control*, 39(8):1564–1574, 1994.

[153] S. Q. Zhu. Graph topology and gap topology for unstable systems. *IEEE Trans. Automat. Control*, 34(8):848–855, 1989.

[154] H. J. Zwart, R. F. Curtain, J. R. Partington, and K. Glover. Partial fraction expansions for delay systems. *Systems Control Lett.*, 10(4):235–243, 1988.

[155] A. Zygmund. *Trigonometric series*. Cambridge University Press, 1988. 2nd edition.

Index